Your Money

in the Age of AI

Boost Income & Take Control Today

D.M. Lewis

First Edition, 2026

ISBN: 979-8-9953379-0-4

Published by: DL Divo Arc Publishing

Disclaimer

This book is intended for educational and informational purposes only and does not constitute financial, legal, or professional advice.

The content, including examples and case studies, is based on publicly available information, general research, and real-world observations. Some details may be simplified for clarity.

While efforts have been made to accurately reflect examples, the author does not guarantee specific incomes, financial results, financial outcomes and individual results may vary depending on effort, experience, and personal circumstances. The author makes no representations or warranties regarding the completeness or reliability of the information presented. The author shall not be held liable for any outcomes, loses, or damages arising from the use or misuse of this material.

Readers are encouraged to seek professional guidance where appropriate before making financial, business, or legal decisions.

Table of Contents

Introduction: Welcome to the Future of Finance and AI

Congratulations on taking the first step towards taking ownership of your finances and wealth-building in the age of artificial intelligence (AI). Before you dive into the strategies outlined in this book, please note that the information provided is for educational purposes and the author does not guarantee any financial or business outcomes from the advice or tools discussed. Please consult a financial professional before making any financial or legal decision.

The world is evolving faster than ever, reshaping how we earn, save, and invest. AI-driven technologies are redefining industries, creating new financial opportunities, and presenting challenges that demand awareness and adaptability.

This book is not about critiquing financial institutions or governing bodies. Instead, my goal is to equip you with the knowledge, understanding and tools needed to navigate this AI environment with confidence as AI, digital currencies, and changing financial policies transform the economic landscape.

How This Book Will Help You

This book is designed to empower you with practical insights so you can take control of your financial future in this era of rapid transformation. Whether you are an entrepreneur, investor, business owner, or simply someone seeking financial security and independence, you will find valuable strategies to help you navigate and capitalize on the AI revolution.

The world is changing fast but with the right knowledge and approach, you can stay informed, prepared, and in control.

What You'll Learn

Beyond understanding AI's impact on finance, this book will introduce you to the power of AI-driven video generation, content creation, and other ways to create new monetization opportunities. Repetition in this work is deliberate; key controls intentionally reappear throughout this book to strengthen awareness and to deepen understanding.

We will explore several areas:

- **AI for Personal, Professional and Financial Growth** – How businesses, educators, and individuals can leverage AI tools.

- **The Future of Money & AI** – How AI is transforming banking, digital currencies, and financial independence.

- **Decentralization & Financial Freedom** – Understanding alternative AI-driven financial systems and protecting wealth.

- **Monetization Strategies** – Turning AI content into income, from YouTube to AI-powered media businesses.

- **Ethical and Legal Considerations** – Navigating AI copyright laws, bias, and misinformation risks.
- **Using AI in Content Creation** – How AI is revolutionizing video production and reshaping creative industries.
- **Digital Banking and Digital Currencies** – their role in financial systems and how to leverage them for financial freedom.

Why This Matters Now More Than Ever

The financial world is experiencing one of the most significant shifts in its history. The rise of AI brings a real possibility of job loss for some but for many it is not a permanent state of unemployment; instead, AI can serve as a turning point, an opportunity to reskill, rethink, and rebuild. Many striving for financial stability can apply practical strategies to adapt and find new pathways to success in an AI- driven world. Also, AI-powered automation, digital banking, and evolving global financial policies are redefining how we interact with money. While these innovations bring greater efficiency and opportunity, they also raise important questions:

- How will AI impact personal finance and investment opportunities?
- What role will digital currencies play in our everyday lives?

- How can we harness these new financial technologies while maintaining our security and independence?

What You'll Discover About AI and Finance

With rapid technological advancements, many people are uncertain about how these changes will affect them personally. That's why there is an urgent need for clear, well-researched, and unbiased information. This book provides insights, practical strategies, and expert advice to help you thrive in the evolving financial landscape.

AI is not just changing how we interact with technology; it is redefining financial freedom itself. This book will serve as your roadmap to navigating and capitalizing on the AI revolution.

Let's dive in!

PART 1

Evolution & Potential of AI in Video Creation

PART 1: Evolution & Potential of AI in Video Creation

Chapter 1: The Early Stages of AI

By understanding the historical context and advancements in AI-driven content creation, you can grasp the full potential of AI video technology and how it continues to reshape industries.

AI video generation refers to the process of using artificial intelligence to create, edit, and enhance video content. AI-driven video production has evolved significantly over recent years. In its initial stages, AI-assisted tools focused on automating small editing tasks, such as background removal and facial recognition for tagging, but presently AI is a doer; it is almost self-sufficient in handling tasks which has resulted in many jobs being replaced.

AI is no longer just a tool; it is a creative partner. From AI-generated scripts to deepfake-powered visuals *(a whole other story because AI should be used responsibly)* video production is evolving at an unprecedented pace. In this chapter, we explore how AI-driven platforms are making high-quality video content accessible to everyone, allowing creators to build brands, generate income, and establish authority online. With advancements in deep learning, AI can now generate entire videos from scratch using scripts, voiceovers, and stock footage.

Key Technological Advancements

1. **Machine Learning & Neural Networks**: Tools that analyze massive datasets to improve video quality and automation.

2. **Natural Language Processing (NLP)**: Enables AI to convert your text into spoken words and video narration.

3. **Computer Vision**: AI that is trained to identify objects, people, and environments, making video editing more precise.

4. **Generative Adversarial Networks (GANs)**: AI can generate hyper-realistic images and videos by learning from real-world footage.

AI-Generated Content in Marketing

Many companies use AI-generated videos for marketing. For example, a major brand, **Coca-Cola** has reportedly utilized AI to create personalized video ads for different audiences, increasing engagement and conversion rates. In the beginning the public did not entirely absorb AI marketing as a warm idea, and there was some minor backlash in how the public perceived this type of marketing. (Forbes – 2024) However, presently many companies are using AI for just that, marketing. AI-powered marketing videos analyze consumer behavior and generate customized advertisements based on user preferences.

PART 1: Evolution & Potential of AI in Video Creation

The Impact of AI on the Film Industry

Hollywood has embraced AI in multiple ways:

- **De-Aging Technology**: AI can modify an actor's appearance to make them look younger (e.g. Martin Scorsese's The Irishman).
- **AI Scriptwriting**: AI can assist in generating script ideas and dialogue.
- **Visual Effects Automation**: AI speeds up CGI and special effects production.

Source: Coca-Cola's AI-Generated Ad Controversy. Explained (Forbes)

Chapter 2: The Rise of AI Influencers and Content Creators

AI video is an innovative way to create content, make money, and build financial independence. It really is a wealth-building tool. The rise of AI in video generation has revolutionized digital content creation by reducing costs, saving time, and increasing accessibility. Traditional video production requires expensive software, professional skills, and considerable time investment. AI-powered tools, however, allow users to generate videos using text prompts, templates, and automated editing features that can be fairly cost effective.

The digital age has transformed how content is created, and AI is leading this revolution. The rise of generative AI tools has enabled individuals, businesses, and educators to create engaging video content faster and more affordably than ever before. Companies like OpenAI, DeepMind, and Google are pioneering

AI research, pushing boundaries on video generation capabilities. AI-driven tools leverage machine learning, deep learning, and neural networks to automate video production, making it easier for individuals and businesses to create high-quality content without extensive technical skills.

Entrepreneurs are now building careers solely around AI-generated videos, whether for YouTube, corporate training, or social media advertising. This section will explore how industries are adapting, job roles are emerging, and the future landscape of AI-powered content.

PART 1: Evolution & Potential of AI in Video Creation

YouTube Entrepreneurs

As mentioned above, many YouTube creators have built six-figure channels with AI-generated content, using tools like **Pictory** and **Synthesia** to generate AI-based videos, including technology reviews that generate income. By consistently uploading high-quality AI-generated content, they can gain thousands of subscribers and monetize their channel through YouTube ads, sponsorships, courses, and affiliate links. By implementing these strategies, other individuals and businesses can generate income using AI-powered video creation tools.

Success Stories in AI Video Content

Synthesia's AI-Powered Corporate Training

Companies are using AI-generated videos for onboarding and compliance training. Synthesia, an AI video platform, has enabled companies like Accenture and Amazon to create employee training videos without hiring actors or production teams.

Case Study: Organizations employing AI-generated videos for employee onboarding, medical presentations, and e-learning content.

Source: "AI Video Case Study Examples: Save Time & Money with Colossyan" by Colossyan. *colossyan.com*

Case Study: Businesses use AI to create engaging e-learning content, enhancing accessibility and engagement

Source: "Case Studies: See how AI videos can save you time & money" by Synthesia. *synthesia.io*

Pictory for YouTube Automation

Pictory helped YouTubers scale their content by automatically converting blog posts into engaging AI-generated videos, reducing content creation time by 70%.

Source: AI-Powered YouTube Automation (TechCrunch, 2024)

Runway ML in Hollywood Productions

AI tools from Runway were used in *Everything Everywhere All at Once* (2022) to assist with visual effects, particularly rotoscoping, significantly reducing the time required for certain editing tasks (Amazon Web Services, 2023; Wikipedia contributors, 2024).

AI-Generated TikTok Success

Influencers are using AI avatars to create viral TikTok videos, automating content creation and driving millions of views with minimal effort. Many of these tools make it easier to even create high-quality video courses without recording yourself, educating people while earning passive income.

- **Choose a Profitable Topic** – Topics like AI, business, marketing, or self-improvement sell well.
- **Use AI to Generate Course Videos** – Tools like Synthesia create talking instructor videos with lifelike AI avatars.

- **Host on Platforms Like Udemy, Teachable, or Gumroad** – Upload your course and set price (e.g., $50 - $200 per student).
- **Promote Your Course on YouTube & social media** – Use short AI-generated clips to drive traffic to your course.

AI-Powered News Reporting

Many organizations are experimenting with AI automated news. Studies show that popular media organizations which may include BBC and Bloomberg experimented with AI-generated news videos, where AI narrated and visualized reports, reducing production costs and improving turnaround time. (MDPI, 2024)

Virtual AI Influencer Campaigns

AI-generated virtual influencers like Lil Miquela, reportedly, collaborated with brands like Prada, earning six-figure sponsorship deals purely from AI-created content. Brands are **paying influencers** to feature their products in AI-generated content. First **secure sponsorship**:

- **Grow an Engaged Audience** – Post regular **AI-generated content** in your niche (e.g., business, finance, tech, beauty).
- **Reach Out to Brands** – Once you have an audience, brands may **pay you $500 - $5,000+ per sponsored video**, although results vary widely depending on scale, clientele, and market positioning.

Use AI to Create High-Quality Ads

AI-generated content is redefining how brands connect with audiences. From explainer videos to product showcases, these tools enable scalable, personalized sponsored content that was previously cost-prohibitive. As highlighted by Vogue Business and NVIDIA, the convergence of AI and media is unlocking new monetization pathways and transforming the economics of digital marketing.

Source: "Finding meaning in the age of AI" by **Vogue Business.** *voguebusiness.com*

Insights: Discussion on how the evolving media landscape, influenced by AI, is creating new revenue opportunities, including sponsored content.

Source: "AI Empowered Monetization: New Content Strategies for Digital Media" by NVIDIA. *YouTube*

AI-Powered E-Learning Videos

AI avatar platforms such as Hour One enable online educators and businesses to create multilingual video content, reducing production costs and supporting global audiences. These tools are increasingly used for training, customer service, and digital learning (Hour One, n.d.).

Case Study: Businesses leveraging AI avatars for multilingual customer assistance and cost-saving content creation. Platforms like Hour One allow organizations to produce training and instructional videos without traditional filming costs, expanding global reach while improving operational efficiency.

Source: Hour One. (n.d.). *AI video case studies*. Hour One

PART 1: Evolution & Potential of AI in Video Creation

AI in E-commerce Product Videos

AI video platforms such as Synthesia are being adopted by e-commerce and enterprise organizations to create product demonstrations, training materials, and multilingual content. These tools significantly reduce video production time and costs while enabling scalable digital engagement (Amazon Web Services, n.d.). *aws.amazon.com*

AI in Real Estate Marketing

Real estate agencies are leveraging AI-generated home tour videos, automating property showcases and saving thousands on professional videographers to market homes in smarter and more affordable ways. This not only saves but also helps stand out online, attract more buyers and sell faster.

What used to require a big budget, and a full team can now be done with just a few clicks giving smaller agencies and independent sellers the power to compete at a much higher level and reach more people than ever before. (Pedra. N.d,)

Source: Pedra. (n.d.). Real estate virtual staging case studies. Retrieved March 30, 2026.

AI-Driven Animated Series

Netflix has demonstrated the business value of artificial intelligence through its recommendation system, which uses machine learning to personalize content and drive user engagement.

Source: Carlos A. Gomez-Uribe & Neil Hunt, 2015).

Once you master these strategies, you can **turn AI video creation into a full-time business**. Here is how:

1. **Automate & Outsource** – Use AI tools to create videos in minutes, and hire freelancers to help with editing, scriptwriting, and posting.

2. **Sell Video Packages & Monthly Subscriptions** – Offer businesses custom AI-generated videos on a retainer basis (possibly $500 - $5,000/month) – these results may vary widely.

3. **Scale with Paid Ads & SEO** – Promote your services using Google & Facebook ads SEO-optimized YouTube videos.

Businesses need promotional videos and product **explainers, and marketing content**, and will pay for high-quality AI-generated videos. Again, it's easy to start….

- **Pick a Niche** – Focus on industries that need videos, like real estate, e-commerce, coaching, and local businesses.

- **Use AI Tools to Create Videos** – Like Synthesia, DeepBrain AI and VideoAi.ai can generate professional business videos.

- **Upsell Monthly Packages** – Sell ongoing content creation video services (e.g., 4 videos per month for $1,000).

- **Sell on Fiverr, Upwork, or Direct Outreach** – Offer custom AI video creation services.

AI enables small teams to deliver high-quality video content on a scale. Matrix Marketing Group noted that AI-driven production workflows are reshaping service models, allowing agencies to compete more efficiently while expanding their revenue potential.

Source: "AI-Driven Case Studies: Streamline Content Creation Heading in 2025" by Matrix Marketing Group. *matrixmarketinggroup.com*

Chapter 3: The Timing and Potential of AI

Is now the best time to leverage AI for income? Look all around you, there is a shift from traditional to AI-generated media. Also, AI is replacing and at the same time creating some new opportunities.

Benefits of AI Video Generation

- **Cost:** Reduces the need for expensive video production teams.
- **Timesaving**: Automates the editing process, allowing faster content production.
- **Scalability**: Enables businesses to create large volumes of content quickly.
- **Accessibility**: Allows anyone, regardless of skill level, to generate professional-looking videos.

Understanding Generative AI Video & Its Potential

Generative AI enables anyone, regardless of technical skill, to create high-quality video content, using machine learning and deep neural networks to generate visuals, animations, and even voiceovers. AI-powered tools can assist in content creation, script writing, voice modulation, and editing.

With the right strategy, these tools can be used for branding, education, training, and entertainment. Understanding the balance between AI assistance and human creativity is key to maximizing this technology's potential. Explore real-world impact of AI-generated films, marketing campaigns, and educational content:

AI's Media Reach

1. **AI in Filmmaking:** The rise of AI-generated short films, and their success in film festivals are increasing. One such short film is Arnaldo Had an Idea which won the **Best International Short** award at the AI Film Festival in Los Angeles, USA.

2. **Educational Content:** AI-generated explainer videos are increasingly being used widely by online course creators like Synthesia. Companies may use a combination of human-led and AI avatars for training.

AI-Powered Ad Campaigns: It is reported that brand name companies such as Coca-Cola and Heinz are using AI videos for marketing purposes to increase engagement and target ads more efficiently

Chapter 4: Creating AI-Generated Videos – A Snapshot

This chapter serves as a practical guide for beginners and experts alike. First things first, not everyone has a big budget, and guess what, that is no longer an obstacle! AI tools now allow content creators to generate professional-grade videos at little to no cost.

This chapter also breaks down a few of the most powerful free and affordable AI tools. It includes a detailed comparison of AI video tools like Synthesia, Pictory, and DeepBrain AI, highlighting their strengths and weaknesses. It will also highlight some tips for crafting compelling narratives using AI-generated scripts to make videos more engaging and impactful videos:

- Best practices for editing and enhancing AI-created visuals to maintain originality and a high-quality look.

- Case studies of brands successfully utilizing AI videos and how they measure success.

- The importance of audience testing, analytics, and iteration to optimize content performance.

PART 1: Evolution & Potential of AI in Video Creation

Use Free & Low-Cost Tools for AI Video Creation

Creating AI-generated videos does not have to be expensive. Many free and low-cost tools allow beginners to create professional-quality videos. These platforms offer essential features such as text-to-video conversion, automated voiceovers, and basic editing capabilities.

Popular AI Video Generation Platforms

- **Synthesia**: Uses AI avatars to create talking-head videos. Companies use tools like Synthesia to create professional training videos, product demos, and customer support guides. AI avatars provide a human-like presence without requiring a film crew, saving time and resources.

- **Pictory**: Converts text-based content into engaging videos.

- **Runway ML**: For AI-powered editing and effects.

- **DeepBrain AI**: Generates realistic AI presenters.

Best Use Cases

- For business/training → Use Synthesia or DeepBrain AI

- For YouTube/social media content → Use Pictory

- For ultra-realistic avatars → Use DeepBrain

Here is a step-by-step example for use:

Pictory.

1. Sign up for a free account on Pictory.ai.
2. Upload a script or blog post.
3. Choose a template and the AI will automatically generate a video.
4. Customize scenes with AI-suggested stock footage.
5. Add voiceovers using AI-generated narration.
6. Download and share your AI-generated video.

By using free tools such as Pictory, you can get started with AI video creation without financial investment.

Best Practices for Creating High-Quality AI Videos

AI can create videos fast, but you need to edit and enhance them to make them look professional. Follow these steps:

Step 1: Craft a Compelling Script

- Keep it short & clear (Aim for 5–7 words per sentence)

- Use storytelling (Make it engage, not robotic)
- Add a hook (Grab attention in the first 5 seconds)

Step 2: Make Your AI Videos Look Professional

- Use background music (Adds emotion & boosts engagement)
- Avoid stock-looking templates (Customize visuals & animations)

Pro Tips:

- Add text overlays & subtitles for better accessibility.
- Choose the right voice and avatar (avoid robotic voices; choose a natural tone) Step 3: Editing & Enhancing AI Visuals
- Even if AI generates the video, you should edit it to make it stand out. Use tools like:
- Canva (For adding effects & branding)
- CapCut (For trimming & transitions)
- Descript (For voice editing & AI cleanup)

Focus on:

- Removing awkward pauses

- Adjusting timing for smooth flow
- Enhancing colors & brightness.

How Brands Can Succeed with AI Videos

Example: A Small Business Growing Sales with AIA clothing brand could use Pictory to create 15-second Instagram ads. By testing different AI-generated scripts and visuals, they could significantly build engagement and boost sales.

Example: YouTuber Scaling with AI - A content creator could utilize Synthesia to make weekly explainer videos on finance topics, without appearing on camera. Within 6 months, they should see significant traction both in subscriber count and ad revenue.

The Power of Analytics & Optimization

Creating AI videos is just the first step. To maximize engagement and profit, you need to test and optimize periodically.

How to Improve Your AI Videos Over Time

- Monitor analytics (Watch time, retention rate, clicks)
- A/B test assorted styles (Try different avatars, voices, and formats)
- Listen to audience feedback (What do they like or dislike?)

PART 1: Evolution & Potential of AI in Video Creation

Example: A brand using AI videos for marketing tested two versions of the same ad. The version with human-like voice AI had 3x more engagement than the robotic voice version.

AI video tools can save time and money, but you still need creativity and editing to stand out.

Start Today:

- Choose the right tool for your needs.
- Write engaging scripts.
- Edit AI visuals for high-quality results.
- Test & optimize for better performance.

With the right approach, AI videos can help you scale your business, YouTube channel, or marketing strategy faster than ever before!

Here are the Top Free AI Video Creation Tools

1. **Canva (with AI features)** – Offers free AI-powered templates for video creation.
2. **Lumen5** – Converts blog posts and articles into engaging AI-generated videos.
3. **Pictory (free tier)** – Allows users to generate short marketing videos from text.
4. **Kapwing** – AI-powered video editing and subtitling

PART 2

AI Ethics, Copyright, and Compliance

Chapter 5: Overcoming Challenges and Ethical Considerations

As AI-generated content becomes more popular, it's important to understand the legal, ethical, and practical challenges that come with it. AI-generated content is powerful but requires responsibility. By understanding copyright laws, ethical AI use, and regulatory compliance, you can create high-quality, legally protected, and fair AI content.

This chapter will provide information on:

- Ensuring fair use and copyright protection
- Ethical frameworks to maintain transparency and responsibility.
- How businesses can address AI bias to foster inclusivity
- AI & copyright laws – how to protect your content legally.
- Addressing AI bias and inclusiveness to create fair and ethical content
- The impact of AI on jobs and industries and how to adapt.
- The ethical responsibility of AI creators in content production.

- Privacy concerns and user-generated AI content, with best practices for safety Navigating misinformation and deepfake risks in an AI-driven world

- Regulatory considerations and compliance for AI video creation

This section will break things down in layman's terms so you can protect your work, stay compliant, and use AI ethically.

1. AI Copyright Laws – How to Protect Your Content Legally

AI-generated content raises questions regarding copyright ownership and fair use.

The Problem

AI can generate images, videos, and text, but who owns the rights to that content? Many AI tools train publicly available data, leading to legal gray areas.

Solutions to Protect Your AI-Generated Work:

1. **Check AI tool policies** – Some platforms claim rights to AI-generated content, so always read the fine print.

2. **Add original elements** – Modify AI-generated content with your own creative touch to strengthen copyright claims.

3. **Trademark & copyright protection** – While AI content alone may not be

copyrightable, logos, branding, and edited AI work can be.

4. **Avoid using copyrighted material** – Never use AI to recreate or copy someone else's work without permission.

Example: If you can, create an AI-generated video, add custom graphics, a voiceover, and unique storytelling to make it legally yours.

2. Addressing AI Bias & Inclusivity – Creating Fair and Ethical Content

The Problem

AI can **inadvertently be biased**, leading to unfair representation or discrimination in content. This happens when AI models are trained on imbalanced data.

Solutions

- **Test your content** – Look for bias in how AI represents diverse groups (race, gender, etc.).
- **Use diverse training data** – If you train AI models, include varied perspectives.
- **Manually review & edit AI outputs** – Don't blindly accept what AI generates, review and adjust as needed.

Example: An AI-generated hiring video that only features male CEOs might be unintentionally biased. Fix this by diversifying representation in the video.

3. The Impact of AI on Jobs & How to Adapt

A broad societal question is: should AI literacy and training be made universally accessible and affordable, given its growing role in survival and everyday life? If yes, who should bear the responsibility of funding and providing that education?

The Problem

AI automates many tasks, and many jobs are at risk of disappearing. Read this whole book to answer these questions, if there is no job, where will the money come from to buy groceries, to buy anything at all?

Solutions

- **Learn AI skills** – Instead of fearing AI, learn to **use it as a tool** to make your work more efficient. Shift towards AI-enhanced careers – Focus on jobs that involve AI management, strategy, or creativity.
- **Adapt business models** – If AI automates something, **offer a service that AI can't fully replace** (like consulting or human-led storytelling).

Example: Instead of losing a job in video production, a professional can offer AI-assisted video editing services to speed up workflows.

4. The Ethical Responsibility of AI Creators in Content Production

The Problem

AI-generated content can be misleading, and creators have a responsibility to use it ethically.

Solutions

- **Label AI-generated content**- Let people know when something is AI-generated to build trust.
- **Avoid misleading edits** – Don't use AI to deceive viewers (e.g., by altering real events).
- **Be transparent about AI use** – Clearly communicate when AI was involved in the content creation process.

Example: If you create an AI-generated documentary, disclose that AI helped with voiceovers or visuals, so viewers are aware.

5. Privacy Concerns & User-Generated AI Content: Best Practices

The Problem

Be very mindful that deep-fake attacks can damage people's reputation. AI can be misused to create realistic voice clones, deep-fake videos, and personal data leaks. Additionally, remember that AI conversations are not stored on your computer or device;

it is stored on someone's server and processed elsewhere. Your data is collected.

Solutions

- Get permission before using someone's likeness. This applies to AI-generated faces, voices, or text.
- Do not assume AI conversations are totally private by default. You are still in control of how much information you divulge. Share only what's necessary.
- Always check your settings/privacy settings. To exclude using your private data unnecessarily.
- Use ethical AI tools – Avoid software that scrapes data without consent.
- Respect personal privacy – Don't use AI to impersonate or manipulate people.

Example: AI-generated customer service voices should be trained on approved voice actors, not cloned from random people without consent.

6. Navigating Misinformation & Deepfake Risks

The Problem

AI can create hyper-realistic fake videos, leading to misinformation or fraudulent content.

Solutions

- **Verify AI-generated information** – Cross-check before sharing AI-created news, facts, or quotes.
- **Use AI responsibly** – Avoid making deep-fakes for deception (e.g., impersonating public figures without consent).
- **Implement content authentication** – Watermark AI-generated videos or include disclaimers.

Example: A deepfake about a politician saying something controversial could spread misinformation. Adding disclaimers or digital watermarks prevents misuse.

Regulatory Considerations & Compliance with AI Video Creation

The Problem

Different countries are introducing AI laws and not following them could lead to fines or legal issues.

Solutions

- **Stay updated on AI regulations** – Laws around deepfakes, copyright, and data protection are evolving.
- **Follow platform rules** – YouTube, TikTok, and Instagram have guidelines on AI-generated content.

- **Use AI tools that comply with regulations** – Choose software that follows laws and has ethical standards.

Example: In the European Union, AI content must follow GDPR rules to protect personal data. Make sure your AI-generated content doesn't violate privacy laws.

AI Content with Ethics & Responsibility

Take Control: Before publishing AI-generated content, **review it for accuracy, bias, and compliance** to ensure it is ethical and legal.

Key Points

- AI-created works may not be eligible for copyright protection, depending on jurisdiction (US Copyright Office, 2023).
- If AI is used as an assistive tool, human creators may claim copyright (WIPO, 2022).
- AI training on copyrighted material can lead to legal disputes (Thompson, 2023).

7. Registration and Other Steps in Protecting Your Content Legally

- **Register Original Content** – If AI-generated content is modified significantly, it may qualify for copyright protection.

- **Use Copyright-Free Sources – Avoid** unauthorized use of copyrighted material in AI training.
- **Read and Abide by the Guidelines** – YouTube, TikTok, and other platforms have evolving guidelines on AI-generated content.

Example: The US Copyright Office ruled in 2023 that AI-generated works without human authorship do not qualify for copyright, but AI-assisted works may be partially eligible (USCO, 2023).

Sources:

- U.S. Copyright Office – Guidelines on AI-generated content copyright (www.copyright.gov)
- General Data Protection Regulation (GDPR) – European Union AI & privacy laws (gdpr.eu)
- World Intellectual Property Organization (WIPO, 2022) AI and intellectual Property Balancing Innovation and Protection
- Deepfake Regulations – U.S. and international rules on AI-generated media (check reputable legal source chapter)
- AI Ethics Studies -Published by MIT, Harvard, and industry leaders on ethical AI practices.

PART 2: AI Ethics, Copyright, and Compliance

Chapter 6: AI Ethics

AI models often reflect the biases present in their training data, raising serious concerns around fairness and representation. These biases may be introduced intentionally (through deliberate design or manipulation) or unintentionally (due to lack of training, lack of diverse data or oversight). Either way, the consequences can be harmful, particularly when AI systems reinforce stereotypes or unintentionally exclude certain groups. This raises a broader societal question: should AI literacy and training be made universally accessible, given its growing role in survival and everyday life? If yes, who should bear the responsibility of funding and providing that education?

Key Challenges:

- AI-generated images and text may reinforce racial, gender, or cultural biases.
- Limited diversity in training datasets leads to skewed, unbalanced AI outputs.
- Bias may arise accidentally through historical data or intentionally through engineered objective.

Solutions for Ethical Content Creation:

- **Diversify Training Data** – Include diverse voices, faces, and languages in datasets.

- **Conduct Bias Audits** – Periodically review AI content to identify and mitigate bias.
- **Encourage User Feedback** – Enable community reporting in apps or content to flag problematic outputs.

Example: Amazon's AI recruiting tool was scrapped after it was found to be biased against women, reflecting historical male-dominated hiring patterns (Dastin, 2018). Although the bias wasn't intentional, it had serious real-world effects, proving that unintentional bias is just as critical to address.

Is AI Evil or Is It Good? A Biblical & Ethical Reflection

Artificial intelligence is one of the most transformative technologies of our time. While some view it as a tool of progress, others fear it as a potential path to global control, moral decay, or even the "mark of the beast" mentioned in the Book of Revelation.

So how do we, as Believers and ethical individuals, approach AI without being misled or misguiding others?

AI itself is not inherently evil, just as electricity, the internet, or fire are not. It is a **tool** created by humans, shaped by the intentions behind its use. Like any tool, it can be used for good or harm.

"Woe to those who call evil good and good evil, who put darkness for light and light for darkness…" **(Isaiah 5:20)**

PART 2: AI Ethics, Copyright, and Compliance

The fear of AI often stems from its potential misuse especially when tied to:

Central Bank Digital Currencies (CBDCs) that may restrict freedom of trade.

Surveillance systems that limit personal privacy.

Algorithms that forget the vulnerable: poor, marginalized or elderly people.

Weaponry that is completely autonomous.

Deepfakes and deception that could blur truth and manipulate the masses. These concerns are valid. The Bible warns us to be watchful in the last days (Matthew 24:4–5). But AI does not have to become the embodiment of control and deception. It can also be part of restoration, justice, and empowerment if used wisely and ethically.

1. AI Can Be Used for Good **Medical Breakthroughs**: AI helps diagnose diseases faster, particularly in rural areas with limited doctors. It can translate health data and support mobile clinics for tribal and forgotten communities.

2. **Education & Literacy**: AI-powered tutors can reach places where schools do not exist. Indigenous children can access tailored lessons in their own languages.

3. **Crisis Management**: AI can be used to forecast natural disasters, such as floods and

earthquakes, helping governments and faith-based relief missions respond swiftly and save lives.

4. **Spiritual and Emotional Support**: Some ministries use AI to translate the Bible into unreached languages or create devotional content for remote believers.

Balancing Faith and Technology in a Challenging World

The real danger is not technology itself, but when humans treat AI as more than it is; when a human-created technology begins to shape our sense of truth, morality or possibly identity, we must examine it thoughtfully. For believers in God, our trust is foundational ("You shall have no other gods before me." (Exodus 20:3)) and that conviction, I believe, calls us to examine AI and to participate in shaping it responsibly and ethically.

Balancing Warning and Hope

It is right to warn against systems of control that could arise from misuse, like the centralization of digital finance or manipulation of the media. At the same time, it is essential not to paralyze people with fear but to arm people with essentials to survive, with the tools and resources at hand.

For those who believe, we are called to be watchful, wise, and hopeful. Just as God used Joseph's wisdom to prepare for famine (Genesis 41), we can use discernment and tech wisely, to teach others in these times.

How to Stay Anchored in Truth

- Pray for discernment and understanding.
- Stay rooted in scripture, not just headlines.
- Use AI tools to educate and uplift—not deceive or exploit.
- Advocate for AI ethics, fairness, and inclusion of the forgotten.
- Support leaders and developers who align with godly, fair values.

AI is not the beast. But its misuse can be a symptom of the systems that may one day support it. That's why God's people must not shrink back but engage the conversation with wisdom and truth. When used with care, the hope is that AI can be used for good and possibly part of God's restoration, even to the nations forgotten by the world.

***"Behold, I am doing a new thing… I will make a way in the wilderness and rivers in the desert."* (Isaiah 43:19).**

AI Impacts on Jobs and Industries – How to Adapt

AI automation is transforming industries like content creation, journalism, marketing, and customer service. These changes are both a challenge and an opportunity.

Industries Affected:

- **Content Creation** – AI tools automate writing, video, and graphic design.

- **Customer Service** – AI chatbots handle queries once managed by humans.
- **Journalism** – AI produces basic news reports, although it lacks the nuance of investigative reporting.
- **Marketing** – Disrupts traditional roles, requiring new skills and at the same time, provides an opportunity by tailoring more personalized preferences using substantial amounts of data to reach the right people at the right time.

How to Adapt:

- **Upskill and Reskill** – Learn to use AI tools in your profession.
- **Prioritize Creativity** – Human ingenuity and storytelling remain unmatched.
- **Human-AI Collaboration** – Combine efficiency of AI with human insight.

Example: Hollywood's 2023 Writers Guild strike included concerns about AI's use in scriptwriting, demanding safeguards for creative control (Baker, 2023).

The Ethical Responsibility of AI Creators in Content Production

As AI-generated media becomes more mainstream, creators and companies hold a critical ethical responsibility to ensure transparency, accuracy, and accountability. AI systems, while powerful, can inadvertently generate content that misinforms, misleads, or amplifies bias, whether through oversight, lack of diverse data, or flaws in design.

Best Practices:

- **Clearly Label AI Content** – AI-generated videos, articles, and visuals should include disclaimers or watermarks to avoid misleading audiences.
- **Ensure Authenticity** – Strive to prevent the spread of AI-generated misinformation by establishing internal fact-checking protocols.
- **Editorial Oversight** – AI-generated journalism and commentary should be reviewed by a human to ensure historical and factual accuracy.

Understanding Intent Matters

AI outputs that mislead or offend aren't always there sult of malicious programming. Bias and misinformation can occur:

- **Intentionally**, through poorly designed prompts or models built to manipulate outcomes.
- **Unintentionally**, due to skewed training data, outdated information, or lack of representation in model design.

Sidebar Insight: ***Intentional vs. Unintentional Bias.***

- ***Intentional bias*** can include discriminatory outputs, unfair targeting, or persuasive content used unethically.

- ***Unintentional bias*** arises when models mirror flawed societal structures, reflect imbalanced datasets, or lack cultural nuance.

Regardless of intent, **impact matters most**, and creators must proactively address harm.

Example: Meta's AI chatbot was criticized for generating historically inaccurate claims (Vincent, 2022). While this was not necessarily intentional, the incident underscores the critical importance of editorial oversight in AI content deployment.

Even tech giants with vast resources are not immune to the risks that come with deploying large language models without robust safety nets. To Meta's credit, the company (and others like OpenAI and Google) have since reportedly invested in internal ethical AI teams and possibly bias mitigation frameworks.

"The journey to responsible AI isn't about blame, it's about building trust through transparency"

PART 2: AI Ethics, Copyright, and Compliance

AI Risks and Compliance Considerations

AI generated media is transforming content creation, but with that power comes a growing ethical responsibility. Developers, platforms, and users alike must ensure that AI-generated content is used transparently, responsibly, and without deception.

Key Concerns

- AI tools can create convincing but false narratives, including historical inaccuracies, deepfakes, and propaganda.
- The absence of editorial oversight in AI-generated journalism and educational content can mislead audiences.

Best Practices for Ethical AI Use

- **Clearly Label AI Content:** Avoid misleading users by clearly indicating when content is generated or assisted by AI.
- **Ensure Authenticity:** AI should be used to enhance truth, not distort it. Avoid training AI models on unverified or biased data sources.
- **Fact-Check AI Journalism:** Always verify the accuracy of AI-generated content, especially news, historical topics, and educational material.
- **Provide Human Oversight**: AI should support, not replace, editorial judgment.

Institutions must assign accountability for content outcomes.

Example: *OpenAI's ChatGPT* often produces confidently incorrect information, known as "hallucinations"—underscoring the importance of human review and prompt engineering to minimize misinformation (OpenAI, 2023).

Navigating AI Misinformation & Deepfake Dangers

Deepfakes and synthetic media can be weaponized to spread falsehoods and manipulate public opinion as well.

How to Combat AI Misinformation:

- **Fact-Check AI Content** – Don't assume outputs are accurate.
- **Use Detection Tools** – Employ software that identifies manipulated content.
- **Promote Public Literacy** – Teach audiences how to identify and verify AI-generated content

Example: In 2024, a deepfake misinformation video of a global leader, Philippine President Ferdinand "Bongbong" Marcos Jr., declaring war went viral before being exposed as fake, causing temporary panic in those that believed it was real. (Reuters, 2024).

PART 2: AI Ethics, Copyright, and Compliance

Regulatory & Compliance Considerations for AI Users

As AI integrates into every aspect of business and society, compliance with legal frameworks is essential.

Common AI Use Cases & Their Compliance Needs:

1. **Resume Builders** – Must protect user data and avoid biased filtering.
2. **Job-Matching Tools** – Should comply with EEO laws.
3. **Interview Coaching AI** – Must disclose how advice is generated.
4. **Freelancing Platforms** – AI-generated works must honor copyright.
5. **Online Education** – AI learning platforms need accreditation.
6. **Business Automation Tools** – Must follow financial and tax regulations.
7. **Customer Support Bots** – Must avoid misleading responses.
8. **Digital Product Creation** – Ensure legal ownership of AI-generated media.
9. **AI Financial Analysis** – Regulated under investment and trading laws.

10. **Mental Health Bots** – Must avoid diagnosing or treating users without human involvement.

Important AI Regulations:

- **EU AI Act (2024)** – Requires transparency and human oversight.
- **US AI Oversight Bills** – Addresses bias, data use, and copyright issues.
- **China's Deepfake Law** – Mandates visible disclosure of synthetic content.

Staying Compliant:

- **Stay Informed** – Monitor evolving legal standards.
- **Disclose AI Content** – Label or watermark synthetic content.
- **Use Ethical Datasets** – Train AI models with legally sourced and diverse data.

Call to Action – Shared Responsibility

Ethical AI is a shared responsibility. Developers, business leaders, creatives, and policymakers must work together to:

- Audit AI systems regularly for bias or misinformation.

- Promote diversity in development teams to reflect broader human perspectives.
- Hold AI accountable by aligning it with social and cultural values.

Sources:

Baker, P. (2023). AI and the Writers Guild Strike: The Fight for Creative Control.

BBC News. (2023). China Mandates Watermarking for AI-Generated Deepfakes.

Bender, E. M., Gebru, T., McMillan-Major, A., & Shmitchell, S. (2021). On the Dangers of Stochastic Parrots: Can Language Models Be Too Big?

Brynjolfsson, E., & McAfee, A. (2022). The Second Machine Age: Work, Progress, and Prosperity in a Time of Brilliant Technologies.

Buolamwini, J., & Gebru, T. (2018). Gender Shades: Intersectional Accuracy Disparities in Commercial Gender Classification.

Carlson, M. (2022). Automating the News: How Algorithms Are Rewriting the Media.

Chesney, R., & Citron, D. (2019). Deepfakes: A Looming Challenge for Privacy, Democracy, and National Security.

Dastin, J. (2018). Amazon Scraps Secret AI Recruiting Tool That Showed Bias Against Women. Reuters.

European Commission. (2024). The EU AI Act and Its Implications.

GDPR. (2022). General Data Protection Regulation Data Protection Rules.

Jobin, A., Ienca, M., & Vayena, E. (2019). The Global Landscape of AI Ethics Guidelines. Nature Machine Intelligence.

Kaplan, A., & Haenlein, M. (2019). Siri, Siri, in My Hand: Who's the Fairest in the Land?

Reuters. (2024). AI Deepfake Causes Global Panic Before Being Debunked.

YouTube. (2023). Updated Guidelines on AI-Generated Content and Monetization.

PART 3

AI Monetization & Business Growth

Chapter 7: Turning AI Content into Income

Monetization is one of the biggest draws of AI-generated video creation. Creating AI-generated content is only half the journey, the challenge is turning that content into real income. This chapter dives into the different methods people can use to make money from their AI-generated content. Learn several ways to make money using AI-generated videos, including ad revenue, affiliate marketing, digital courses, and freelance services.

Whether you want to make passive money through YouTube, sell AI-generated videos, or build an online brand, this guide will walk you through step by step on how to monetize your content effectively, ways to turn content into cash flow.

Turning AI-generated videos into a money-making business is possible but you need to act. Whether you monetize YouTube, sell videos, create courses, or run an AI content agency, the opportunity is huge!

How to Start Your Own AI-Powered Media Business

- Identify a niche (e.g., educational content, entertainment, marketing videos, explainer videos, religious content, etc.).

- Choose an AI video platform (e.g., InVideo AI, RunwayML Synthesia) or build a custom AI solution.
- Decide on a business model (subscription-based, ad revenue, client services, etc.).
- Set up a website, YouTube channel, and social media for content distribution.
- Leverage automation tools for scaling content creation.

Awareness plays a significant role since technology and tools may change rapidly; observe how businesses and media companies are using AI for large-scale productions. Learn how AI integrates into major streaming platforms and what the future holds for generative content.

Case Study: Several major AI companies, including OpenAI and Google, have reportedly paid content creators for access to unused or unpublished video footage to help train generative AI models.

Source: Counts, Aisha "YouTubers are Selling Their Unused Video Footage to AI Companies." By Bloomberg, 10 Jan 2025 – Https://www.bloomberg.com/news/articles/2025-01-10/youtubers-are-selling-their -unused-video-footage-to-ai-companies.

Insights: Utilizing AI for content creation, media buying, data analysis, and campaign personalization to enhance YouTube marketing strategies.

Source: "AI-enhanced YouTube marketing: Insights from 3 case studies" by Search Engine Land. searchengineland.com

Start small, there are many monetization examples, pick one monetization and commit to testing it for 30-60 days. AI can help in automating processing:

- **Find High-Paying Affiliate Programs** – Join platforms like Amazon Associates, ClickBank, and ShareASale to find products to promote.
- **Use AI to Create Engaging Review Videos** – Make videos comparing products, giving tips, or explaining features.
- **Post on YouTube, TikTok, and Instagram** – Place affiliate links in your video descriptions so people can buy directly; earn commissions on sales.

PART 4

AI for Personal Growth, Therapy & Career Success

Chapter 8: AI Expansion and the Future

Metaverse is like a virtual world or a digital universe, where people can interact, work, play, and socialize using the internet. The metaverse can be thought of as an advanced version of today's online games but with you in the game, virtual reality (VR), and augmented reality (AR), all combined into one massive, interconnected space.

Imagine putting on VR glasses and walking through a virtual city, shopping in digital stores, attending concerts+, or even working in a 3D office, all without leaving your home. AI is helping to build and improve these experiences by making virtual worlds more interactive, lifelike, and personalized.

This chapter will highlight how AI has already transformed how we create and consume content, and how it is shaping the future of digital worlds where people can live, work, and play in new ways in which people can hang out in, attend virtual concerts or meet in #D worlds. Another example is buying, selling and trading digital real estate. The metaverse can confuse the line between online life and reality or offline life.

Expanding Beyond YouTube – Netflix, Amazon Prime, & More

While YouTube is a great platform for creators, breaking into larger or additional streaming services can increase monetization opportunities, a broader audience, and professional recognition.

PART 4: AI for Personal Growth, Therapy & Career Success

This section is about expanding beyond YouTube and leveraging bigger streaming platforms like Netflix, Amazon Prime, Hulu, and other OTT (Over-The-Top) services to distribute AI-generated content.

To successfully expand into these platforms, creators should focus on researching Submission Guidelines. Every streaming platform has its own rules for content submission.

- Netflix, for example, reportedly works with independent filmmakers but often requires partnerships with approved production companies.

- Amazon Prime Video allows direct uploads via Prime Video Direct, but content must meet technical and legal standards.

Creating High-Quality AI-Generated Content

AI can be used to generate documentaries, short films, animated content, or even full-length movies.

- If using AI, ensure professional-grade editing, storytelling, and compliance with streaming platform standards.

Networking & Attending Industry Events

- Engage with film and tech professionals to find opportunities and partnerships.

- Attend events such as Sundance Film Festival, Cannes, SXSW, and AI-specific tech expos to showcase AI-driven content.

Using Crowdfunding

Platforms like Kickstarter, Indiegogo, or GoFundMe can help fund AI-generated film projects.

- A strong marketing campaign can attract investors, film enthusiasts, and early adopters. Exploring Licensing Deals & OTT Partnerships.
- Rather than just uploading content, creators can negotiate licensing deals, where streaming services pay for the rights to host their content.
- Other options include renting or selling AI-generated content on platforms like Vimeo on Demand, Tubi, or Apple TV.

The Role of AI in Metaverse and Digital Storytelling

AI is the engine that powers immersive storytelling through VR/AR environments because it can generate such realistic environments and create virtual characters such as avatars and interactive, real-time narratives. Producers can generate great ideas for a story while AI-generates the in-depth, detailed scripts, voiceovers, and animations which can create hyper-personalized digital experiences. In previous chapters we covered businesses who have used AI for marketing; Businesses can use AI storytelling for marketing in virtual worlds.

PART 4: AI for Personal Growth, Therapy & Career Success

How AI Can Be a Tool for Social Good and Empowerment

Socially, AI can be used to make educational content more accessible, especially for underprivileged communities in which distance to a physical institution can be a challenge. AI-powered storytelling can bridge the gap of soft delivery to raise awareness of social issues and in many instances feel more personalized for each user based on a game, a concert, or an interactive learning experience.

Although AI does not replace human care, AI-based mental health therapy videos and emotional support tools like Chatbot and virtual assistants can provide immediate support, answer questions and guide users through mindfulness exercises, making support available to more people anytime and anywhere.

Similarly, AI-driven accessibility tools are transforming accessibility for people with disabilities, like turning spoken words into real-life captions. AI-driven accessibility can describe images for visually/hearing impaired people, therefore giving everyone a better chance to interact fully.

Building an AI Content Agency and Working with Clients.

Building an AI content agency means helping businesses or creators use AI tools efficiently by offering AI-powered video services, including explainer videos, corporate training, social media ads. Working with clients involves listening and building a portfolio specific to their goals using AI-generated content.

This is a combination of human vision and the AI boost of that vision.

People can network on freelancing platforms like Upwork, Fiverr and LinkedIn to not only highlight their skills and collaborating but also by offering their services such as ai-generated videos, graphics, writing connecting to clients world-wide and gain income, which helps clients and agencies grow faster.

Both creators and businesses have saved time by automating repetitive tasks, for example, batch processing lets you produce multiple videos at once using the same script, style or template without manually editing each one. Template-based AI video creation allows plug-in text, images or voiceovers and instantly generate polished videos for marketing, tutorials or social media content.

AI-Generated Video – The Future is Now!

AI-generated videos are involving fast, with fully AI-generated feature-length films and interactive AI-generated customized content, characters or scenes powered by voice assistants. As AI becomes more powerful, ethical questions are increasingly important. AI-powered technology may introduce additional regulatory restrictions which may drive compliance for ethical storytelling to prevent bias and protect privacy when using personal data to train AI.

Addressing these issues is critical to ensure AI is used responsibly and benefits creators, audiences and society fairly; building trust where humans are viewed as priority and at the same time, benefit from the present AI-driven technology, for instance: AI -driven real-time animation and motion capture can benefit from manually animating every movement to, instead, allowing creators to produce animated films, games or virtual experiences more quickly.

Blockchain can work hand in hand with AI to protect a creator's work and manage their digital rights, for instance, by recording ownership and usage on a ledger, their AI content is tracked and licensed for safe monetization of their content.

AI Video Platforms - Resources

- **InVideo AI** – Quick AI-powered video creation for content marketing.
- **RunwayML** – Cutting-edge AI tools for video editing and generation.
- **Synthesia** – AI avatar-based video creation for businesses.
- **Pika Labs** – AI-powered animation and motion graphics.

Best Online Courses for Mastering AI Video

- **Deep Learning Specialization** – Coursera (Andrew Ng)
- **Generative AI with Runway** – Runway Academy
- **AI Video Editing and Scripting** – Udemy
- **AI for Digital Storytelling** – MIT Open Course Ware

Books, Podcasts, and Thought Leaders in AI

- **Books**: "Artificial Intelligence: A Guide for Thinking Humans" (Melanie Mitchell), "The Creativity Code" (Marcus du Sautoy).
- **Podcasts**: "AI Alignment Podcast," "The AI-Breakdown."
- **Thought Leaders**: Andrej Karpathy, Yann LeCun, Demis Hassabis, OpenAI researchers.

Key Takeaways to Act Now

- Pick a niche and target audience.
- Choose an AI video platform and learn how to use it.
- Create and publish content daily (YouTube, TikTok, Instagram Reels).
- Monetize via YouTube ads, affiliate marketing, and brand deals.
- Scale by offering your services to businesses and influencers

AI can help automate the process:

- Find High-Paying Affiliate Programs – Join platforms like Amazon Associates, ClickBank, and ShareASale to find products to promote.

PART 4: AI for Personal Growth, Therapy & Career Success

- Use AI to Create Engaging Review Videos – Make videos comparing products, giving tips, or explaining features.
- Post on YouTube, TikTok, and Instagram – Place affiliate links in your video descriptions so people can buy directly.
- Earn Commissions on Every Sale – Some product sales may increase commissions.

Chapter 9: Real-World Uses- Education – Therapy- Growth

AI tools can also be leveraged for education, therapy, and business growth. From AI-generated learning modules to mental health support through AI-driven counseling, this section explores real-world applications that improve lives.

AI-Driven Learning

Specifically, AI-driven video tools are revolutionizing education by personalizing learning experiences, automating content creation, and enhancing student engagement. Online education is booming, and AI tools allow creators to generate instructor-led training videos, animated explainers, and interactive lessons. Platforms such as Kajabi, Udemy, and Skillshare enable creators to monetize AI-powered courses.

Key Benefits:

- **Personalized Learning:** AI can adapt videos to different learning levels.
- **Automated Content Generation:** Teachers can create lesson videos quickly using AI-powered tools.
- **Language Translation & Accessibility:** AI-generated subtitles help students with disabilities or language barriers.

Example: Khan Academy's AI tutor helps students receive instant explanations tailored to their needs (Khan Academy, 2023).

Using AI-Generated Content for Therapy and Mental Well-Being

AI-powered mental health applications are now providing accessible support through chatbots, virtual therapists, and personalized self-care plans.

How AI Supports Mental Health:

- **AI Chatbots for Counseling:** Virtual assistants like Woebot and Wysa offer cognitive behavioral therapy techniques (Fitzpatrick et al., 2017).
- **AI-Powered Journaling & Mood Tracking:** Apps analyze user inputs to detect emotional patterns.
- **Meditation & Stress Management:** AI-generated music and guided meditations help users manage anxiety.
- Applications like ***Calm*** provide AI-generated guided meditation and relaxing soundscapes.
- **The Headspace** app focuses on mindfulness and stress reduction with mediation sessions.
- **Endel** focuses on generation personalized soundscapes for sleep, focus, and relaxation.

Example: Woebot, an AI chatbot, showed significant improvements in reducing anxiety and depression symptoms in users (Darcy et al., 2020).

How AI Can Help Businesses Cut Costs and Maximize Profit

AI streamlines operations, improves efficiency, and reduces costs across a variety of industries.

AI Applications for Businesses:

- **Automated Customer Service:** AI-powered chatbots reduce operational costs (Brynjolfsson & McAfee, 2022).
- **Data-Driven Decision Making:** AI systems can analyze business trends and customer behavior for smarter strategies (Davenport & Ronanki, 2018).
- **Process Automation:** AI automates repetitive task such as manual data entry, reducing labor costs (Manyika et al., 2017).

Example: Netflix uses AI to recommend content for users, reducing churn rates and boosting customer retention (Gomez-Uribe & Hunt, 2016).

Real-World Applications of AI – Changing Lives and Businesses

AI technology is empowering individuals to achieve career growth, improve education, and launch businesses.

- **Freelancer & AI Video Editing:** Many content creators report that AI editing tools can reduce video production time significantly by about 60–70%, allowing creators to publish videos much faster than before.

- **AI in Job Hunting:** Career experts and job seekers may agree that AI-assisted resume optimization which improves formatting and optimizing keywords can help job seekers improve their visibility in recruiter reviews and lead to more interviews.

- **AI in Entrepreneurship:** Small business owners, who may not have the budget to hire a marketing team, can turn to AI-generated ad copy tools to help create ads, instead of spending weeks writing copy and designing graphics.

AI-Driven Personal Branding and Career Growth

AI is transforming how professionals build their personal brands and expand career opportunities.

How AI Helps with Personal Branding:

- **AI-Generated Content Creation:** AI automates blog writing and video scripts, for efficient content creation (Clark & Leung, 2023).

- **Resume & LinkedIn Optimization:** AI is able to tailor resumes to specific job descriptions (Patel et al., 2021).

- **Personalized Learning Paths:** AI can suggest courses and skills based on a person's career goals, providing an individual learning program. (Anderson, 2022).

Example: Indeed provides an AI-tailored resume for

users' approval. Similarly, AI-powered LinkedIn profile optimizers increased job search success rates by 35% (Patel et al., 2021).

AI's Role in Self-Improvement and Skill-Building

AI helps individuals develop new skills through personalized learning experiences and adaptive training programs.

Self-Improvement:

- **Language Learning:** Apps like Duolingo use AI to provide tailored lessons based on user progress (Garcia & Lin, 2019).

- **Skill Development:** AI suggests career-relevant skills and training materials to improve professionals' knowledge (Ng et al., 2020).

- **AI-Powered Coaching:** Virtual AI mentors provide feedback on communication and leadership skills (Brown, 2021).

PART 4: AI for Personal Growth, Therapy & Career Success

Example: **AI**-powered learning platforms boosted skill acquisition rates by 50% compared to traditional methods (Ng et al., 2020).

Sources:

Fitzpatrick, K., et al. (2017). Evaluating the Effectiveness of Woebot, an AI Chatbot for Mental Health Support.

Huang, M.-H., & Rust, R. T. (2021). Artificial Intelligence in Mental Health: Opportunities and Challenges.

Goyal, M., et al. (2014). Meditation Programs for Psychological Stress and Well-Being: A Systematic Review.

Brynjolfsson, E., & McAfee, A. (2022). The Business of AI: How Artificial Intelligence Can Transform Companies.

Davenport, T., & Ronanki, R. (2018). Artificial Intelligence for the Real World. Harvard Business Review.

Manyika, J., et al. (2017). Harnessing AI for Business Growth and Efficiency.

Clark, A., & Leung, C. (2023). How AI Is Revolutionizing Personal Branding Strategies.

Patel, R., et al. (2021). AI-Driven Job Hunting: How Algorithms Improve Resume Visibility.

Anderson, K. (2022). The Future of Career Growth with AI Learning Platforms.

Luckin, R., et al. (2018). AI in Education: The Challenges and Opportunities of AI in Learning.

Holmes, W., et al. (2019). Artificial Intelligence in Schools: Promises and Challenges.

Nguyen, T., et al. (2021). The Role of AI in Enhancing Accessibility in Education.

Garcia, M., & Lin, P. (2019). The Effectiveness of AI-Driven language Learning Apps.

Ng, A., et al. (2020). AI-Powered Learning: How Artificial Intelligence Improves Skill Development.

Brown, S. (2021). AI-Powered Coaching: How Virtual Mentors Enhance Professional Growth.

PART 4: AI for Personal Growth, Therapy & Career Success

Chapter 10: Kickstart Your 30-Day AI Content Challenge

This **30-Day AI Content Challenge** can help individuals or businesses develop their AI-powered content creation skills and kickstart their journey into the world of AI-generated content. It can progressively build skills and create valuable output. It is also designed to help you explore the diverse ways in which AI can be leveraged to create high-quality, engaging content, written articles, videos and social media posts, and integrate AI into your content creation business or personal projects.

Day 1–5: AI Writing Fundamentals

Goal: Familiarize yourself with AI tools for writing content.

Day 1: Introduction to AI Writing Tools

- Try AI writing assistants like **ChatGPT** or **Jasper AI** for generating articles, blog posts, or even social media captions. +

- Write and refine a simple 500-word blog post or article n a topic of your choice using AI.

Day 2: Create a Product Description

- Use AI to generate a detailed and persuasive product description.

- Incorporate keywords to make the description SEO-friendly.

Day 3: Write an Email Campaign

- Create a three-email series for a marketing campaign using AI. Include a welcome email, promotional mail, and follow-up email.

Day 4: Create Social Media Posts

- Use AI to generate five engaging posts (including text and hashtags) for platforms like Instagram, LinkedIn, or Twitter.

Day 5: Generate a Blog Post Outline

- Use AI to generate an outline for a long-form blog post (800-1,500.) text and hashtags) for platforms like Instagram, LinkedIn, or Twitter.

- Make the outline SEO-optimized with headings and subheadings.

Day 6–10: AI for Visual Content

Goal: Learn how to use AI for creating images and videos.

Day 6: Create a Social Media Graphic

- Use AI tools like **Canva** or **Fotor AI** to create an eye-catching graphic for a social media post.

Day 7: Generate a Logo Design

- Use AI-driven logo design tools like **Looka** or **Hatchful** to generate a professional logo for your brand or business.

PART 4: AI for Personal Growth, Therapy & Career Success

Day 8: Create an AI-Generated Video Intro

- Use tools like **Pictory** or **Synthesia** to create a 15–30 second AI-generated video intro for your content or brand.

Day 9: AI Image Enhancements

- Use AI-based tools (like **DeepArt** or **Luminar AI**) to enhance and edit images, improving their quality and appeal

Day 10: Video Editing with AI

- Use AI-based video editors like **Magisto** or **Descript** to edit and optimize a one-minute video (e.g., a promotional or explainer video).

Day 11-15: Advanced AI Writing and Content Strategy

Goal: Build more advanced and targeted content using AI.

Day 11: Write a Long-Form Blog Post

- Use AI to write a 1,500–2,000-word blog post on a complex topic, ensuring it's SEO-optimized and informative.

Day 12: AI-Generated Story or Script

- Use AI tools to generate a story or script for a short video, commercial, or creative project (300–500 words).

Day 13: SEO Content Generation

- Use AI to generate SEO-optimized blog posts, including keyword research using tools like **Ahrefs** or **SEMrush**.

Day 14: Create an E-Book Outline

- Use AI to generate an outline for a 10–20-page e-book and start developing content for the first chapter.

Day 15: AI Content Strategy Planning

- Use AI to develop a 30-day content strategy plan for social media, blog posts, or email marketing.

Day 16-20: AI Content for Social Media and Marketing

Goal: Leverage AI for targeted social media and marketing content.

Day 16: Create an AI-Powered Influencer Pitch

- Use AI to generate a pitch email for a brand partnership or influencer collaboration.

Day 17: Write AI-Generated Ad Copy

- Create a series of AI-generated ads for Facebook or Google Ads, including targeting keywords and calls to action.

PART 4: AI for Personal Growth, Therapy & Career Success

Day 18: Develop a Marketing Funnel Using AI

- Use AI tools to generate a complete marketing funnel (landing pages, email series, ads) to guide customers toward conversion on your website.

Day 19: Generate AI-Enhanced Testimonials or Case Studies

- Write a realistic testimonial or case study, showcasing customer satisfaction or product success.

Day 20: AI-Generated Video for Marketing

- Create a short AI-generated promotional video, utilizing text-to-video tools like **Pictory** or **Lumen5.**

Day 21-25: AI for Data, Research, and Personalization

Goal: Use AI for research, data-driven content, and personalization.

Day 21: AI-Powered Research

- Use AI tools like **Quillbot** or **ChatGPT** for summarizing articles and extracting key insights for blog posts.

Day 22: Content Personalization Using AI

- Create a personalized email or landing page experience using AI-driven content generation, addressing individual user preferences.

Day 23: AI for Market Research

- Use AI tools to analyze competitors, market trends, or customer behavior, creating data-driven content.

Day 24: Develop Personalized Recommendations

- Use AI to create personalized content recommendations based on user behavior (e.g., recommended articles, products, or videos).

Day 25: Generate Infographics with AI

- Use AI tools like **Visme** or **Venngage** to create infographics summarizing key data or trends.

Day 26-30: AI for Video and Interactive Content

Goal: Create advanced video content and interactive AI experiences.

Day 26: Create a Webinar Presentation

- Use AI to develop a compelling webinar script and PowerPoint slides, utilizing AI insights and data.

Day 27: Interactive Chatbot Content

- Use AI-powered chatbot tools to create an interactive FAQ page or conversational interface for a website.

PART 4: AI for Personal Growth, Therapy & Career Success

Day 28: Automate Content Scheduling

- Use AI tools like **Buffer** or **Hootsuite** to schedule and optimize content posting across multiple platforms.

Day 29: Create a Two-Minute Explainer Video

- Use AI tools to create a two-minute explainer video explaining your product, service, or brand message.

Day 30: AI-Generated Content Review

- Review all your AI-generated content over the past month. Use AI tools to improve, optimize, or repurpose previous content into new formats (e.g., turning blog posts into social media snippets or videos).

Tips for Your - AI 30-Day Content Challenge

- **Tracking Progress:** Keep a journal of what you learn each day and experiment with new AI tools and platforms.

- Consistency is Key: Try to follow the challenge daily to ensure continuous improvement and learn the diverse applications of AI for content creation.

- **Networking:** Share your AI-generated content on social media and engage with others in the AI and content creation community.

Chapter 11: AI Learning for Every Budget Online Courses (Free)

Learn AI starting with zero dollars to $100+. The best approach for some is to combine free online learning, and paid learning resources based on budget and goals.

Fast.ai – Practical Deep Learning for Coders *(Beginner-Friendly)* https://course.fast.ai/. Hands-on deep learning course with PyTorch and real-world applications. Learn how to code; said to be used by people with various levels of skillsets.

Harvard's CS50 AI – Free AI Course *(Beginner-Friendly)* https://cs50.ai/. Covers foundational AI concepts, search algorithms, and neural networks.

Google AI – Learn with Google AI *(Beginner & Intermediate)*

https://ai.google/education/ Tutorials on machine learning (ML), TensorFlow, and responsible AI.

AWS: Start free; allows hands on experience with Amazon's services - **https://skillbuilder.aws**

IBM SkillsBuild: Provides AI skills for free. **https://skillsbuild.org**

OpenAI: Learn with AI experts, webinars, etc. **https://academy.openai.com**

PART 4: AI for Personal Growth, Therapy & Career Success

Elements of AI (University of Helsinki & Reaktor) *(Beginner)*

https://www.elementsofai.com/ Introductory AI course covering ethics, machine learning, and neural networks.

MIT OpenCourseWare – Artificial intelligence *(Intermediate)*

https://ocw.mit.edu/courses/electrical-engineering-and-computer-science/6-034-artificial-intelligence-fall-2010/

Covers AI problem-solving, machine learning, and natural language processing.

YouTube Channels for AI Learning (Free)

- **DeepLearningAI** (by Andrew Ng) – Practical ML and AI career advice. https://www.youtube.com/c/Deeplearningai
- **3Blue1Brown** (Math for AI & ML) – Explains AI mathematics in an intuitive way: https://www.youtube.com/c/3blue1brown
- **Two Minute Papers** (AI Research Made Easy) – Summarizes cutting-edge AI research in simple terms. https://www.youtube.com/c/KárolyZsolnai
- **Sentdex** (Python AI/ML Tutorials) – Hands-on Python tutorials for AI

https://www.youtube.com/user/sentdex

AI Coding & Research Platforms (Free)

- **Google Colab** (Run Python & AI Models in the Cloud for Free) https://colab.research.google.com/
- **Hugging Face** – Free AI Models & NLP Research: https://huggingface.co/
- **Kaggle** – Free AI Datasets & Competitions: https://www.kaggle.com/

Free AI Learning Platforms:

- **YouTube Channels:** Two Minute Papers, AI Coffee Break with Letitia, Yannic Kilcher.
- **Communities**: The Subreddit r/MachineLearning, AI Discord groups, Twitter AI influencers.
- **Fast.AI** – Provides free deep-learning. Learn how to code. This tool is known to be used by people with different skillsets.

Try these free AI Books:

- *Deep Learning* by Ian Goodfellow, Yoshua Bengio, and Aaron Courville (Advanced); https://www.deeplearningbook.org/ Theoretical and practical concepts in deep learning.

PART 4: AI for Personal Growth, Therapy & Career Success

- Neural Networks and Deep Learning by Michael Nielsen (Intermediate): Covers backpropagation, neural networks, and implementation. http://neuralnetworksanddeeplearning.com/

- Reinforcement Learning: An Introduction (by Richard Sutton & Andrew Barto) (Advanced). https://www.andrew.cmu.edu/course/10-703/textbook/BartoSutton.pdf A deep dive into reinforcement learning (AI used in robotics & gaming).

Low-Cost Books and Links to Learn AI (As of 2025)

Budget $50 and under….

To increase AI knowledge, stay ahead by investing time and researching the best resource for your learning. Perplexity AI is good for research and has a free version with limits.

AI Books Under $50.

- Your Money in the Age of AI – Boost Your Income & Take Control Today (Check Amazon)

- "AI Superpowers" by Kai-Fu Lee (Global AI Industry Trends) ~$15 (https://amzn.to/3JCWJGj)

- "The Hundred-Page Machine Learning Book" by Andriy Burkov (Beginner-Friendly) ~$30 (https://amzn.to/3IAD8wj)

- "Hands-On Machine Learning with Scikit-Learn, Keras, and TensorFlow" by Aurélien Géron (Best for Python AI Programming) ~$45 https://amzn.to/3Tsq3Iw)

- "AI Superpowers" by Kai-Fu-Lee.

- "The 100-Page ML Book" by Andriy Burkov.

Low-Cost AI Online Courses $50 and under (Paid & Budget-Friendly)

- **Coursera** – AI for Everyone (by Andrew Ng) Free to Audit, Certificate: ~$49 https://www.coursera.org/learn/ai-for-everyone

- **Udemy** – AI & Machine Learning Courses $10-$20 (Frequent Discounts) https://www.udemy.com/courses/search/?q=artificial%20intelligence. Beginner to advanced courses in AI, ML, and automation.

- **DataCamp** – AI & Data Science Training $25/month (First Chapter Free) https://www.datacamp.com/ Interactive coding for AI, Python, and data science.

Harvard's Data Science & AI Courses (edX)

Free to Audit, Certificate: ~$99+ https://online-learning.harvard.edu/course/cs50s-introduction-artificial-intelligence-python

- **Udacity** – Combination of free and paid courses.

Budget $100+: AI Mastery (Subscription & Premium)

For those looking to invest in professional AI education, consider an AI specialization certificate.

- OpenAI API & GPT-4 Subscription (For AI Development) $20/month (ChatGPT Plus) https://openai.com/
- DeepLearning.AI Specialization (Coursera) $49/month https://www.deeplearning.ai/
- MIT AI Professional Certificate $2,000 (Installments Available)

  https://executive.mit.edu/course/artificial-intelligence-implications-for-business-strategy/
- ISACA - AAIA™ Certification | ISACA Advanced in AI Audit™
- IAPP - AI governance | IAPP

More Resources and Tools to Get Started Today

- **AI Video Generators**: RunwayML, Pika Labs, In-Video AI, Synthesia, HeyGen.
- **AI Voiceovers**: ElevenLabs, Murf.ai, Play.ht.
- **Script Generators**: ChatGPT, Jasper AI, Copy.ai.
- **Editing Tools**: CapCut, DaVinci Resolve, Adobe Premiere Pro with AI plugins.

AI YouTube Channels That Took Off: It is safe to say that there are many YouTube channels that are successful, especially those that share AI knowledge, capability, tools and general content. YouTubers can earn a substantial amount of money per month, on several topics including health and celebrity news and finance, with millions or hundreds of thousands of subscribers as well as sponsors.

Example: Influencers monetizing unused video content by selling it to companies training AI, generating significant income.

Source: "Influencers are making big money selling leftover videos - ones not yet posted online - to train AI" by The New York Post. nypost.com

Freelancers Earning from AI Video Services

Because there is a great demand to speed up manual tasks, or efficiently complete tasks like cutting and trimming video footage or just enhancing video quality or producing AI-generated voice overs, freelancers can increase earnings from AI video services.

PART 4: AI for Personal Growth, Therapy & Career Success

Case Study: Companies utilizing AI video creation platforms to significantly reduce production time and costs, leading to increased revenue.

Source "Synthesia Makes AI Video Production Effortless with Generative AI" by Amazon Web Services - *aws.amazon.com*.

Action Plans

As you may know by now, AI-generated videos can **earn additional income** through ad revenue, sponsorships, and memberships on platforms like YouTube, TikTok, Instagram, and Facebook. **Here are additional keys to success:**

1. **Create & Post AI Videos Consistently** – Use tools like Synthesia, Pictory, or DeepBrain AI to generate engaging videos. Focus on popular topics like tech, finance, motivation, or tutorials to attract viewers.

2. **Join YouTube's Partner Program (YPP)** – To qualify, however, I believe you need:

    - 1,000 subscribers

    - 4,000 watch hours in the past year or 10 million views on Shorts

3. **Enable Ads on Your Videos** – Once approved, YouTube will start running ads on your content, and you'll earn a share of the ad revenue. Do your research on this.

4. **Use Super Chats, Memberships & Exclusive Content** – Platforms like YouTube allow fans to donate, subscribe for exclusive perks, or tip your content directly.

Startups can use AI Video for Growth: Many companies leverage AI videos in marketing directly to consumers, training videos to educate their viewers or subscribers as well.

PART 5

AI & Financial Freedom

Chapter 12: Protecting Your Money with AI Apps

The banking sector is influenced by government policies for compliance purposes and with changes in regulations, interest rates, inflation control, digital currencies can impact how people save, invest, and access their money.

If not affected already, here's how these changes might affect the banking sector in the future:

Regulations & Increased Oversight

- Expect more banking regulations aimed at preventing economic crises, increasing security, and protecting consumers from unfair fees.

- Tighter regulations on cryptocurrency and digital assets to control money laundering and fraud.

- Expansion of Central Bank Digital Currencies (CBDCs), which could increase transparency but also government control over transactions. (CBDCs are discussed in later chapters.)

Inflation & Interest Rate Changes

- Policies that raise or lower interest rates affect loan accessibility, mortgage rates, and returns on your savings accounts.

- If inflation continues to rise, banks may increase fees and limit withdrawals, making it harder for people to access their money freely. This makes sense since your money in the Bank is essentially used for investment purposes.

Bank Failures & Bailouts

- Some smaller banks may collapse due to debt and mismanagement, leading to government bailouts.

- The government may enforce stricter bank reserve requirements, limiting how much banks can lend.

Digital Banking & Less Cash Use

- The push for cashless transactions and AI-driven banking could make it easier to track and tax every transaction.

- People may lose privacy as the government has more access to financial data.

Protect Your Money, Take Control

Banks operate as profit-driven businesses, often charging high fees for transactions, maintenance, and other services. In today's economic climate, where every dollar counts, AI-powered financial tools offer individuals greater control over their money. From automated budgeting and fraud detection to AI-driven investment strategies and decentralized finance (DeFi), these innovations provide smarter, cost-effective alternatives to traditional banking.

By leveraging AI, individuals can enhance their financial security, reduce reliance on costly banking services, and explore new ways to grow their wealth in an ever-changing financial and technological landscape.

AI Tools for Personal Finance & Wealth Protection

1. **AI Budgeting Apps** – Apps like YNAB, Mint, or Cleo use AI to optimize spending, avoid unnecessary bank fees, and automate savings.

2. **AI Investment Bots** – Platforms like Wealthfront & Betterment invest money smartly with AI, reducing risk and helping people make passive income.

3. **Decentralized AI Banking** – AI-driven DeFi (decentralized finance) systems let people earn interest outside of traditional banks, such as in decentralized lending platforms in which platforms may not be branded AI but

use AI risk tools, such as Aave, Compound, MakerDAO (via Oasis.app).

On the AAVE platform, users lend their cryptocurrency to others and earn interest or passive income without a traditional bank. AI determines if the borrower qualifies for a loan using risk models and fair interest rates.

Another example is where AI bots are used to analyze market conditions, even past behaviors or trends and execute trades automatically. Users may earn without actively managing their portfolio, using platforms like 3Commas, Zignaly and SnapsotX/Gelato Network.

4. **AI Fraud Protection** – Tools like Darktrace AI & IBM Watson monitor bank accounts in real time to detect fraudulent transactions.

5. **AI-Powered Tax Planning** – AI tools like TurboTax AI optimize tax returns and prevent overpayments to the IRS.

AI-Driven Tools for Financial Independence

1. **AI Investment Assistants** – Platforms like Betterment, Wealthfront, and Magnifi AI help users build passive investment portfolios with ease.

2. **AI Budgeting & Finance Apps** – Apps like Cleo and Copilot AI provide real-time

financial insights and eliminate unnecessary banking fees.

3. **AI Credit & Loan Optimizers** – Tools like Upstart and Tala offer personalized, lower-interest loan options outside of traditional banks.

4. **AI Tax Optimization** – Smart tax assistants like FlyFin AI help maximize deductions and reduce taxable income.

Bottom Line: AI is also enhancing financial freedom, making it easier for individuals to manage, grow, and protect their wealth, without excessive fees or restrictions.

Reducing Poverty: AI & Alternative Strategies

Money management can be difficult, especially when traditional banks charge fees, offer high-interest loans, or require good credit to access certain financial services. Many people struggle to save or grow their money because they don't have access to the right tools or knowledge.

AI is making financial help available to everyone. For example, there are AI-powered apps which track spending, suggest ways to save money, and even help people invest tiny amounts over time. Some apps round up spare change from purchases and automatically put it into savings pots, while others create smart budgets to help people avoid overspending.

Alternative banking options, like digital wallets and online financial services, also offer lower fees and easier access to money even for those without a traditional bank account. AI can also help people find better loan options, avoid scams, and understand how to build credit without getting trapped in debt.

By using these AI-driven tools, people from all backgrounds, whether students, working professionals, or retirees, can better manage their money, reduce financial stress, and build a more secure future.

Use AI for Passive Income & Side Hustles

- **AI-powered drop-shipping,** print-on-demand, automated video creation, and freelance platforms allow people to generate income with minimal upfront costs.
- **AI-driven crypto trading bots** help users invest wisely without being experts in trading.
- **AI monetization platforms** (YouTube, TikTok, AI-generated content) create automated revenue streams.

Get Paid in Decentralized Digital Currency (Crypto, DeFi)

- Avoid bank fees & inflation by saving in Bitcoin, Ethereum, or stablecoins instead of traditional bank accounts.
- Use AI-powered DeFi lending & staking to earn interest without using banks.

- AI-based P2P lending platforms allow people to lend & borrow without banks charging extreme interest rates.

AI-Based Credit & Loan Alternatives

- AI-driven lending systems like Kiva & Upstart provide low-interest loans for small businesses without needing a traditional bank.AI-powered microfinance platforms help individuals in poor communities access capital.

AI-Powered Financial Literacy Education

- AI tutors & chatbots (like Khan Academy AI & ChatGPT) teach people how to invest, save, and avoid financial scams.

- AI-driven stock trading simulators let people learn to invest without real money risk.

AI & Tech Changing the Game

Traditional banks have long been the primary place to store, manage, and borrow money. While they offer convenience and security, they also rely on fees, interest rates, and strict requirements that can sometimes make financial growth challenging.

Today, AI-powered tools and decentralized finance (DeFi) provide new opportunities for people to manage their money in smarter, more flexible ways. These innovations allow individuals to automate savings, access lower-cost financial services, and explore alternative ways to grow their wealth without becoming fully dependent on traditional banks. By understanding these options, people can make informed choices that best fit their financial goals and lifestyle.

Smart Ways to Manage Your Money with AI & Alternative Banking

Keep Only What You Need in Traditional Banks

- Store just enough for regular expenses while considering alternative savings options, like gold, digital assets, or on decentralized financial platforms.
- Use AI-powered savings tools to automatically move extra funds into higher-yield accounts rather than letting them sit in low-interest bank savings accounts.

Use AI to Spot Hidden Bank Fees

- AI-driven expense tracking apps can alert you to unexpected bank fees, high transaction charges, and overdraft penalties.
- Consider switching to online banks like Chime or Ally, which often have fewer fees and better rewards.

Automate Savings Beyond Traditional Banks

- AI-powered banking apps like Acorns allow you to round up spare change from purchases and invest it automatically.

- Smart investing tools can help grow your money without requiring constant monitoring or large upfront deposits.

Explore AI-Powered Crypto & DeFi Tools

- Decentralized finance (DeFi) platforms like AAVE and Compound use AI to automate lending and earning opportunities, often without traditional banking fees.

- AI can also help track crypto investments and analyze market trends, giving you insights into smarter financial decisions.

Use AI for Legal & Financial Protection

- AI-driven legal tools like DoNotPay can help dispute unfair banking fees, overdraft charges, and financial mismanagement.

- AI-generated reports can provide clear evidence of wrongful charges, making it easier to contest fees and protect your rights.

Chapter 13: Controls and Alternatives to Financial Freedom

As technology evolves and digital finance becomes more mainstream, individuals now have more control over their financial future than ever before. AI is revolutionizing how people manage, grow, and protect their money, offering smarter, more accessible alternatives to traditional banking.

Traditional banking has long been the foundation of finance, but fees, debt, and limited access can often hold people back. AI is shifting the landscape by providing more efficient ways to:

- Monitor and manage finances with smart budgeting tools.
- Automate wealth-building strategies without needing financial expertise.
- Invest wisely through data-driven decision-making.
- Protect assets with advanced security and fraud prevention.

Useful Apps to Take Control Today

You can use AI-powered financial apps like **CLEO** to track spending, **Emma** to connect all bank accounts in one place, and **Mint** which is a budgeting tool, to track spending and income and to avoid hidden fees.

- Explore AI-driven side hustles and automated income sources to create financial stability.
- Diversify savings through alternative assets like digital currencies, gold, and decentralized finance (DeFi).
- Automate tax planning and fraud detection with AI-powered security tools.

By embracing these AI-driven financial strategies, individuals can increase their financial independence and create lasting wealth in a rapidly changing world.

Future-Proofing Your Financial Freedom Using AI

Since banks, AI, and global economies are shifting rapidly, now is the time to take smart financial steps.

Steps to Financial Freedom

1. **Automate Wealth Growth** – Use AI-powered investment platforms to build your money passively.
2. **Go Cashless but Stay Decentralized** Use AI-driven DeFi apps to store and invest wealth without traditional banks.
3. **Protect Your Financial Privacy** – Use AI-powered privacy tools to avoid overreach & surveillance.

4. **Use AI to Start Side Hustles** – AI can generate content, automate drop-shipping, and create digital businesses for passive income.

5. **Diversify Wealth Outside Banks** – Invest in crypto, AI-driven trading, DeFi lending, and alternative assets.

Bottom Line: The future belongs to those who embrace AI-driven financial independence NOW.

Is AI the Key to Financial Empowerment?

We are witnessing a once-in-a-lifetime economic shift where AI and decentralization are taking power away from traditional banks. Those who adapt early will build AI-powered wealth and financial independence. Those who remain dependent on banks may face tighter control, higher fees, and economic instability.

Act Now:

- Use AI for smart investing, budgeting, and passive income.
- Explore DeFi, smart contracts, and AI-powered wealth tools.
- Secure financial independence before banks tighten their grip further.

AI and decentralization are here, present and future. Will you control financial security?

The Evolution of Banking & Financial Systems

As AI and decentralized finance (DeFi) expand, financial institutions are adapting to these new innovations. Governments and banks are exploring digital banking solutions and introducing policies that aim to balance financial security with innovation.

Some key trends include:

1. Increased regulation on digital banking and crypto.

2. Exploration of Central Bank Digital Currencies (CBDCs) as new forms of digital transactions.

3. Greater use of AI for financial security and fraud detection.

What does this mean to you? These changes highlight the growing importance of financial literacy and personal control over your wealth. By staying informed and leveraging AI-driven financial tools, you can make the most of emerging opportunities.

Exploring AI-Powered Financial Alternatives

More people than ever before are exploring alternative financial systems that offer greater flexibility and security. A couple of options noted before are emphasized below:

1. **Decentralized Finance (DeFi)** – Platforms like **AAVE** and **Compound** allow

individuals to borrow, lend, and earn interest without traditional banks.

2. **AI-Powered Crowdfunding** – Platforms like **GoFundMe AI** and **Patreon AI** help individuals and businesses secure funding for projects and investments.

3. **Smart Contract-Based Transactions** – Blockchain technology, such as **Ethereum, has** smart contracts, which enables secure, automated financial transactions without intermediaries. These Smart Contracts may be written in languages like Solidity.

4. **AI-Driven Investing** – AI-powered platforms like **Numerai** and AI-driven hedge funds help users make data-driven investment decisions.

5. **AI in Real Estate Investing** – Tools like **Lofty AI** allow people to invest in real estate and earn passive income without requiring large capital.

Bottom Line: The financial landscape is evolving, and AI is opening doors to new, accessible, and smarter ways to manage money. By adopting AI-powered financial tools, individuals can create a stronger, more resilient financial future.

Final Thoughts

The future of money is AI-enhanced, decentralized, and built for individual financial empowerment. While traditional banking will continue to play a role, AI-driven financial tools give people more choices, more control, and more opportunities to achieve financial independence. By staying informed and embracing new technologies, anyone can take charge of their financial future.

Chapter 14: AI & The Global Economy: What's Next?

Am sure you can agree that AI is not just transforming banking; it's changing the entire global economy.

AI Will Replace Traditional Finance Jobs

- Bank tellers, loan officers, and financial advisors are already being replaced by AI-powered financial assistants.
- Robo-advisors and AI stock trading bots may have outperformed human financial managers in tests.

AI and The End of Paper Money

- Central authorities worldwide are transitioning toward digital-only currencies (CBDCs).
- AI-powered tax systems will track and control financial transactions more easily.

Chapter 15: Universal Basic Income (UBI) and Impact

As AI replaces more jobs, central authority might introduce **AI-driven Universal Basic Income (UBI) systems** to distribute money fairly. This means that instead of people relying solely on wages, that central authority would provide a set amount of money to citizens regularly, ensuring financial stability. Can you imagine?

AI-driven UBI systems would likely use AI algorithms to determine eligibility, distribution, and fairness, making sure funds are allocated efficiently. This idea has been widely discussed in academic papers, economic reports, and books on the future of work and AI.

Understanding AI-Driven Universal Basic Income (UBI)

Also, as AI and automation take over more jobs, many experts believe governments may need to introduce **a UBI** system. This means that instead of relying only on wages, people would receive a fixed amount of money regularly (e.g., monthly) from the government to help cover basic needs.

PART 5: AI & Financial Freedom

Why Would We Need Universal Basic Income (UBI)?

- **AI is Replacing Jobs** – Automation is making many traditional jobs disappear, especially in industries like manufacturing, retail, customer service, and even creative fields.
- **People Still Need Money** – If jobs are scarce, people will need another way to afford food, housing, and healthcare.
- **Keeps the Economy Stable** – If people have money to spend, businesses keep running, and the economy doesn't collapse.

How Would AI Help Manage UBI?

AI could play a role in making UBI fair and efficient by:

- **Tracking Employment Trends** – AI can analyze job markets to determine which industries are most affected by automation.
- **Determining Who Needs UBI** – AI could help decide who qualifies for UBI based on financial need.
- **Preventing Fraud** – AI can detect suspicious activities and ensure funds go to the right people.

- **Customizing Payments** – Some AI models may adjust UBI payments based on factors like cost of living in different areas.

UBI – Pros and Cons

UBI payments may be based on factors like cost of living in different areas

Pros	Cons
• Helps people afford necessities even if jobs disappear	• Expensive for governments to fund
• Reduces poverty and economic inequality	• AI decisions may be biased or unfair
• Encourages entrepreneurship (people can take risks without fear of losing income)	• Some people may lose motivation to work

• Keeps the economy running by ensuring people have spending money	• Risk of increased government control over finances

Real-World UBI Experiments

Governments and organizations have already tested UBI in small-scale programs, including:

- **Finland (2017–2018)**: Gave 2,000 unemployed people €560 per month, no strings attached. It improved well, but didn't significantly boost employment.
- **Stockton, California (2019–2021)**: Provided $500 per month to 125 low-income residents. Many used it for food, rent, and medical expenses but there were concerns about dependency on government aid.
- **Canada (Ontario's Pilot Program)**: Launched a trial in 2017 but was canceled early due to political reasons.

What's Next?

It is believed by some technology leaders, like Elon Musk and Mark Zuckerberg, that concepts such as Universal Basic Income (UBI) may need to be explored as a potential response to job displacement caused by AI and automation. However, critics argue that such systems may be difficult to sustain without significant funding mechanisms, such as increased taxation on automation or large corporations.

Emerging discussions around artificial intelligence, automation, and economic transformation suggest that new models of income distribution, such as Universal Basic Income (UBI) and guaranteed income programs, could reshape how people earn and access financial resources. While these models remain debated, they are increasingly explored as potential responses to job displacement and structural shifts in the global economy.

Research from The Second Machine Age, Rise of the Robots, and the World Economic Forum highlights how automation and AI are transforming labor markets, while The War on Normal People discusses UBI as one possible solution.

Real-world experiments provide additional insight.

Finland's basic income experiment (2017–2018), led by the Finnish government, found improvements in well-being and modest employment effects. Similarly, the Stockton, California guaranteed income pilot (2019–2021) showed that recipients experienced increased financial stability and, in some cases, higher employment rates.

While the long-term sustainability of such programs remains under debate, these studies suggest that alternative income models may play a role in future economic systems shaped by AI and automation. Below are sources and case studies into the respective basic income experiments and their outcomes:

Sources:

Brynjolfsson, E., & McAfee, A. (2014). *The Second Machine Age: Work, Progress, and Prosperity in a Time of*

Brilliant Technologies. W.W. Norton & Company.

Ford, M. (2015). *Rise of the Robots: Technology and the Threat of a Jobless Future*. Basic Books.

World Economic Forum. (2020). *The Future of Jobs report.*

Yang, A. (2018). *The War on Normal People*. Hachette

Case Studies:

Kangas, O., et al. (2019). *The Basic Income Experiment*

2017–2018 in Finland: Preliminary Results. Finnish Government.

NPR. (2021). *California program giving $500 no-strings-attached stipends pays off, study finds*.

CBS News. (2021). *Study shows Stockton universal basic income experiment increased employment*

PART 6

The Rise of CBDCs: Control & Privacy Risks

Chapter 16: How CBDCs Differ from Cryptocurrency

CBDCs (Central Bank Digital Currencies) are not like Bitcoin or other decentralized cryptocurrencies. Reportedly, CBDCs are being developed by major governments and central banks worldwide, including the U.S. Federal Reserve, the European Central Bank, and the People's Bank of China.

These digital currencies may seem like a modern financial innovation, but many are concerned that CBDCs come with hidden dangers that could possibly give governments unprecedented control over people's money and lives. Unlike cash, which allows private, untraceable transactions, CBDCs are issued and managed directly by central banks, meaning:

- Central authorities can track every dollar you spend
- Your money could be frozen or programmed with restrictions
- Spending could be controlled based on policies or social criteria
- Cash elimination could force 100% reliance on digital transactions

Just as AI is revolutionizing global content, AI is also being used by governments and financial institutions to reshape the global economy. Central Bank Digital Currencies (CBDCs) are being promoted as the future of money, but they come with, what some call, serious risks. Unlike decentralized cryptocurrencies, CBDCs could give central authorities the ability to track, limit, or even freeze personal finances at will. Do not confuse CBDCs with decentralized cryptocurrencies like Bitcoin, they are fundamentally different:

Cryptocurrencies vs CBDCs

Feature	Cryptocurrencies	CBDCs
Control	Decentralized (No central authority)	Government-controlled
Privacy	Transactions can be private	Full surveillance by authorities
Ownership	You own your crypto	Government can freeze or delete funds
Supply	Limited (e.g., 21M Bitcoin max)	Unlimited (governments print as much as they want)
Censorship	Cannot be blocked	Transactions can be restricted or reversed

CBDCs allow governments and banks to monitor and control money in real time, making **financial freedom challenging.**

What's in Place and What Led Up to CBDC?

1. The Federal Reserve is Already Exploring a Digital Dollar

- The FedNow instant payment system was launched in 2023, laying the groundwork for a full CBDC system.
- It's reported that both administrations have studied CBDCs under the Executive Order on Digital Assets.
- The U.S. Treasury and Federal Reserve have both signaled they want CBDCs to modernize the financial system.

2. Financial Surveillance is Increasing

- Banks already report transactions over $600 via the IRS.
- AI-powered banking surveillance flags "suspicious transactions," sometimes freezing accounts with no due process.
- The U.S. Patriot Act expanded financial monitoring, and new AI tools are being used to track crypto transactions.

3. Privacy Laws Are Weak, and We Need Help

- Big Tech shares personal financial data with government agencies, often without congressional oversight.

PART 6: The Rise of CBDCs: Control & Privacy Risks

- Recent government actions may show a trend toward increased monitoring of financial transactions, even without public approval.

- Opposition to CBDCs exists, but laws protecting cash and financial privacy are not being prioritized.

How CBDCs Can Be Used for Control

A. Programmable Money: Conditional Spending in Real Time

As central banks race toward launching their own CBDCs, there's growing concern about how this new form of money could be used not just as a financial tool, but as a mechanism of control.

CBDCs may offer convenience, but they also come with risks that could reshape the relationship between citizens and the state. CBDCs allow governments to program money with rules, controlling how, when, and where you spend.

Below are some of the most pressing implications:

- **Spending Limits**: Money could be blocked from being used on "unapproved" items. For example, if you've exceeded your **carbon emissions quota**, you could be blocked from buying gas or booking flights.

- **Expiration Dates**: Funds could expire after a set time, nudging consumers to spend quickly removing the concept of long-term saving.

- **Geo-Fencing**: CBDC usage could be restricted to specific regions. Funds meant for "local economic stimulation" might only work within your city or state.

B. Negative Interest Rates: Your Money Loses Value

One of the features central banks are exploring is the ability to apply negative interest rates directly to your CBDC balance. That means your money could shrink over time if you don't spend it fast enough. This penalizes saving and could make long-term wealth-building nearly impossible.

It would also push people into perpetual consumption and dependency on the state's economic system.

C. Freezing and Seizing Funds: Instant Asset Control

CBDCs give governments direct access to your digital wallet. That means they could freeze or seize funds in real time, without needing traditional banking intermediaries.

- Participating in a protest deemed "unlawful"? Your account could be frozen.

- Donated to a politically sensitive organization? Your funds might be flagged or withheld.

- Failed to comply with vaccine mandates, tax filings, or ID verification? Expect restrictions.

Example: In Canada (2022), the government froze bank accounts of peaceful protesters during the Freedom Convoy demonstrations. A CBDC would've made this process even faster and harder to avoid.

D. UBI with Strings Attached: Explore Dependency Over Independence

Universal Basic Income (UBI) is gaining traction globally. But when tied to CBDCs, UBI risks becoming a digital leash rather than a lifeline.

- Funds might expire each month if unused.
- Spending could be limited (no travel, alcohol, or luxury purchases).
- Continued access could require behavior compliance—such as attending mandatory programs, completing surveys, or maintaining a digital ID score.

This doesn't empower citizens; it creates a modern welfare trap.

E. Social Credit & Financial Surveillance

In countries like The People's Republic of China, it's reported that **social credit systems** already tie behavior to privileges, limiting travel, job opportunities, and loans based on "trustworthiness." With CBDCs, this level of surveillance becomes easier to scale.

If a social score is linked to your CBDC wallet, the consequences could include:

- Reduced access to credit or housing.

- Inability to purchase flights or trains.
- Denial of insurance, business licenses, or even education opportunities.

In China, citizens can be banned from flying or using high-speed trains based on their social score. This was noted in prior chapters.

CBDCs may be framed as innovation but their programmable, traceable, and centrally controlled nature makes them a double-edged sword. In contrast to decentralized cryptocurrencies that offer freedom and anonymity, CBDCs offer convenience at the cost of control.

Most governments around the world, whether democratic or authoritarian, regardless of political stance, will likely adopt some version of CBDCs within the next decade. The real question is: Will they use this technology to empower citizens or to restrict them?

Options for Protecting Your Finance

- **Diversify Wealth:** Store assets in multiple forms, including gold, Bitcoin, and privacy-focused crypto **(e.g., Monero).**
- **Use Self-Custody Wallets:** Don't keep large amounts of money in banks. Use hardware wallets (**Ledger, Trezor**) to store crypto securely.
- **Support Decentralized Finance (DeFi)**: Platforms like Uniswap, AAVE, and

PancakeSwap let you earn and trade without banks.

- **Use Privacy Coins**: Monero and Zcash provide private transactions that can't be tracked easily.
- **Invest in Real Assets:** Land, food supplies, and barterable goods hold value outside government-controlled money.

Educate Others: The more people who reject CBDCs, the harder it will be for governments to enforce them.

AI-powered financial tracking tools are now capable of analyzing billions of transactions in real-time, detecting patterns, and flagging any behavior that regulators deem suspicious. But what happens when governments use these tools to control spending, enforce social credit systems, or restrict financial access?

Summary – What to Expect If CBDCs Are Implemented

CBDCs are marketed as a convenient, digital form of money, but they represent a dangerous shift in financial control. Once fully implemented, reversing their effects will be nearly impossible. The time to prepare is now by learning about decentralized alternatives and spreading awareness.

AI-powered financial tracking tools are now capable of analyzing billions of transactions in real-time, detecting patterns, and flagging any behavior that regulators deem suspicious. But what happens when governments use these tools to control spending, enforce social credit systems, or restrict financial access? If the CBDC system is launched, it could mean:

1. **Transaction Monitoring** – Every purchase, donation, and financial move would be visible to government agencies?

2. **Spending Restrictions** – Money programmed so that it expires, is restricted in certain industries, or is limited for individuals based on regulations?

3. **Freeze & Seizure Risks** – If your views, actions, or spending patterns are deemed **"problematic,"** your funds could be frozen or reduced?

4. **Negative Interest Rates** – Government could force people to spend their money by imposing negative interest rates, reducing savings over time?

5. **Cash Elimination** – A move toward full digital control means cash could be phased out entirely, leaving no anonymous transactions.

PART 6: The Rise of CBDCs: Control & Privacy Risks

Chapter 17: AI, Decentralization & Financial Privacy

What are Your Options for Privacy?

Protect Your Financial Privacy with AI-Powered Tools

1. Avoid overreach & surveillance by using AI-powered privacy solutions.
2. Privacy coins like Monero (XMR) and Zcash (ZEC) allow anonymous transactions.
3. AI-enhanced privacy wallets like Samourai Wallet & Wasabi Wallet protect Bitcoin transactions.
4. Decentralized VPNs (dVPNs) like **Orchid & Mysterium** encrypt online activity.

Wealth Outside the Banking System

Don't keep all your assets in a financial institution, which can be controlled by an entity.

1. **Gold & Silver**: Physical assets hold value regardless of digital currency changes.
2. **Bitcoin & Cryptocurrencies**: Decentralized, censorship-resistant money.
3. **Real Estate & Hard Assets**: Tangible wealth that can't be erased digitally.

The Cashless Shift: Keep Your Freedom

Are CBDCs already happening? Yes, globally CBDCs are already active in some countries, and CBDCs are in development or pilot stages in other places like Asia, Brazil (DREX) and Russia. Given the increasing AI-driven surveillance of financial transactions, it's clear that implementation and control over digital money is becoming a reality.

- The question is not IF CBDCs will be implemented, but WHEN will everyone adopt CBDCs?
- Will Americans or nations resist, or will they accept a fully controlled financial system?
- Will CBDCs become the future of money as we know it now?

Challenges of Cashless Transactions via Traditional Systems

Many governments and banks arc pushing for a cashless society, but their version of it comes with:

1. Full transaction monitoring & tracking
2. Frozen accounts & spending restrictions
3. Higher banking fees & loss of financial autonomy

4. The risk of being de-platformed from the financial system CBDCs will give central authorities full control to

5. Limit what people can spend money on

6. Freeze funds instantly (if they deem your actions suspicious)

Impose expiration dates on money (forcing people to spend or lose it).

Example: China's digital yuan **(CBDC)** already has spending restrictions, and Europe & the U.S. are exploring similar systems.

The Future of Money: CBDCs, Government Control & AI-Driven Financial Freedom

AI-Powered DeFi as the Solution

Instead of relying on banks or any entity issued digital currencies, AI-driven DeFi apps allow you to go cashless while keeping control of your wealth.

Key AI-Powered DeFi Tools:

- **Crypto wallets** – Store wealth outside banks using AI-powered wallets like MetaMask AI, Trust Wallet AI, or Ledger Nano AI.

- **DeFi lending & borrowing** – Use AAVE, Compound, or MakerDAO to lend & borrow funds without banks.

- **AI-automated DeFi trading** – Use AI-powered bots like 3Commas or Pionex to automate crypto trading and yield farming.
- **Smart contracts** – Set up automated financial agreements without middlemen.

Result: Control, direct your wealth and operate in a cashless world.

The Stealthy Rollout of CBDCs: How Banks Are Quietly Preparing Without Public Attention

While central banks and governments publicly claim that CBDCs are only in the "research phase," behind the scenes, they are already laying the groundwork for a full rollout. Here's how banks may be discreetly preparing CBDCs.

Cashless Evidence: Gradual Removal of Cash & ATMs

While digital payments are convenient, they also come with trade-offs. Around the world, the use of physical cash is declining as digital payment systems become more common. As this shift continues, it raises important questions about financial access, privacy, and control.

Let's explore the potential risks of a more cashless economy and how individuals can protect their financial independence. Some smaller or regional banks have reduced physical branch locations or merged with larger institutions, which can also lead to fewer ATMs in certain areas.

PART 6: The Rise of CBDCs: Control & Privacy Risks

- In 2023, the collapse of s Silicon Valley Bank, Signature Bank, and First Republic Bank experienced high-profile failures, increasing public concerns about the stability of some portions of the banking system.

- Major banks like Bank of America, JPMorgan Chase, and Wells Fargo have adjusted policies over time, including withdrawal limits and branch services, although these vary by account and location.

Some businesses have moved toward cashless payments for convenience, efficiency, or health considerations, especially following the COVID-19 pandemic. Systems like FedNow, launched in 2023, enable faster digital payments. While separate from digital currencies, they reflect a broader trend toward real-time, digital financial infrastructure. Government responses to banking crises have included interventions to stabilize the financial system, which can sometimes benefit larger institutions due to their systemic importance.

Why this matters: As cash becomes less common, more financial activity moves into digital systems. This creates benefits in speed and convenience, but also raises ongoing discussions about privacy, access, and financial independence.

What you can do: As the financial system continues to evolve, individuals can take practical steps to stay prepared:

- **Keep multiple payment options** – Maintain access to both digital and physical forms of money where possible.

- **Diversify where you store money** – Avoid relying on a single bank or financial platform.
- **Stay informed** – Pay attention to changes in banking policies, digital payment systems, and financial regulations.
- **Maintain emergency access** – Ensure you have access to funds during outages, restrictions, or unexpected disruptions.
- **Understand modern technologies** – Learn how systems like FedNow and digital currencies work so you're not caught off guard.

Being prepared does not mean avoiding progress, it means understanding it and making informed decisions.

Phasing in "Digital Wallets" & AI-Driven Banking Controls

- Banks are rolling out "centralized digital wallets" under the guise of convenience.
- AI-powered transaction monitoring is increasing, flagging people's spending patterns.
- Some bank accounts may have already been frozen under vague "risk assessment" rules.

It is believed that PayPal updated its terms to allow $2,500 fines for "misinformation", imagine what CBDCs could do under full government control.

Why This Matters: Digital wallets and AI-driven

restrictions pave the way for programmable money that can be controlled remotely.

Preparing the Public for "Programmable Money"

Organizations like the International Monetary Fund and the Bank for International Settlements have been exploring digital currencies issued by governments, often called CBDCs (Central Bank Digital Currencies).

One of the key ideas behind these systems is something called "programmable money." This means, in theory, money could be designed to behave in certain ways, for example, being used only for specific purposes, or even having time limits when governments issue financial support.

While these features are still being discussed and are not fully implemented in most places, they have raised important questions. Some people believe this could help governments deliver aid more efficiently. Others are concerned it could lead to more control over how money is used.

Events like the temporary freezing of bank accounts during protests in Canada have also made people more aware of how access to money can be affected by policy decisions, even though that situation did not involve digital currency.

Why this matters: As financial systems evolve, there is an ongoing conversation about balance, between innovation and convenience on one side, and personal financial freedom and privacy on the other. However, many are still pursuing crypto wallets as a solution; users can still access and move funds freely, even across borders.

Chapter 18: Is Your Local Bank Disappearing?

Let's talk about centralized banking. As the financial landscape rapidly evolves, there's a noticeable trend flying under the radar: local and regional banks seem to be closing or consolidating, while larger centralized financial institutions and government-influenced digital systems rise in dominance. This shift may pose a serious risk to community-level financial independence and increase our dependence on centralized systems, particularly CBDCs.

What's Happening?

- **Branch Closures Are Rising**: Thousands of local bank branches have closed in recent years, especially in rural and lower-income areas. This cuts off easy access to physical banking and encourages reliance on digital platforms.

- **Consolidation of Banks**: Smaller banks are being absorbed by larger institutions or forced to shut down due to rising compliance costs and changing tech demands. This limits consumer choices and increases financial surveillance.

- **Digital Push Accelerates**: As more financial services go digital, people without access to broadband, smartphones, or digital literacy are at risk of being left behind.

- **CBDC Rollout Pressure**: Reportedly, governments and central banks around the world are preparing infrastructure for digital currencies, which may eventually replace physical cash. This change centralizes control of financial data, possibly allowing for programmable restrictions on spending.

Why Bank Closures Matter

Local and regional banks serve specific community needs and offer a personalized service, flexible lending, and financial inclusion. Their disappearance could:

- Increase reliance on national/global financial institutions that have tighter controls.
- Diminish access to funding for small businesses and community development.
- Shift power toward a few major banks or government entities, enabling tighter oversight over individual financial behavior.

What You Can Do

- **Diversify Your Financial Tools:** Use a mix of traditional banks, DeFi platforms, and community credit unions when possible.

- **Support Community Banks:** Keep accounts at local banks or credit unions that invest back into your community.

- **Learn Digital Finance Safely:** Use decentralized apps and privacy-focused tools that align with your values.

Stay Informed: Watch financial policy shifts in your country especially those related to cash withdrawal limits, CBDC rollouts, or the closure of small bank branches. These are often early indicators of a shift toward centralization and control over individual financial freedom.

Sources:

Bis.org. (2023). Central Bank Digital Currencies: A New Era of Control? https://www.bis.org

CNBC. (2023). Bank Closures Continue, Especially in Rural America. https://www.cnbc.com

Federal Reserve Bank. (2023). Trends in Branch Banking.

https://www.federalreserve.gov/publications.htm

FDIC. (2022). Community Bank Study. https://www.fdic.gov/resources/community-banking/report/

IMF & World Economic Forum. (2022). The Future of CBDCs and Cross-Border Payments.

https://www.imf.org https://www.weforum.org

PART 6: The Rise of CBDCs: Control & Privacy Risks

Is AI-Driven Financial Surveillance Already Here?

- Banks and payment processors are already using AI- driven "suspicious activity" monitoring.
- Some people are getting their bank accounts frozen or shut down for vague "policy violations."
- Crypto exchanges are facing increasing government restrictions, making it harder to exit the traditional financial system

Why This Matters: If traditional accounts can be frozen now, imagine how much easier it will be with CBDCs controlled at the central bank level.

Are DeFi Banks & AI-Powered Financial Platforms Insured?

1. DeFi platforms are NOT FDIC insured because they are not controlled by central banks or the U.S. government.
2. Some crypto-based banks offer insurance through private entities, but they are not federally backed.
3. Smart contract risks & hacking are the biggest threats in-DeFi.

Safer Options in Decentralized Finance:

- Bitcoin (BTC) self-custody wallets (e.g., Ledger, Trezor)
- Privacy coins like Monero (XMR)
- DeFi lending platforms like AAVE, Compound (smart contract risks exist)
- Decentralized exchanges (DEXs) like Uniswap & PancakeSwap

"Hybrid" Banking Options (With Private Insurance):

- Proton Bank (crypto-friendly with privacy protection)
- Swan Bitcoin (Bitcoin-focused, no CBDC risk)
- Custodia Bank (Wyoming-based, Bitcoin-focused, pushing for federal approval)

How to Stay Ahead of the CBDC Rollout

Diversify assets into decentralized options (gold, silver, Bitcoin, privacy coins)

- Use self-custody wallets & hardware wallets to protect digital wealth
- Limit dependence on big banks & centralized financial institutions

- Educate others on financial freedom before it's too late

Chapter 19: Is this a Race for Your Money?

Is this a financial takeover? While CBDCs are being sold as "just an idea," the reality is that the infrastructure is already being set up. The goal is to make the transition seem "natural" and "inevitable." By the time people realize what's happening, it may be too late to opt out. Stay aware, take control, if not some, and control your financial future before being controlled.

Why Decentralized Finance (DeFi) Matters

DeFi platforms powered by blockchain and supported by AI can offer individuals greater financial autonomy, reducing the risk of asset freezes, restrictions, or surveillance.

Five Steps to Safeguard Your Financial Freedom:

1. **Self-Custody Your Assets**: Use hardware wallets like Ledger or Trezor to store your cryptocurrencies offline, protecting them from hacks and centralized interference.

2. **Engage with Decentralized Platforms**: Trade or invest using DEXs (Decentralized Exchanges) like Uniswap or PancakeSwap, which are not controlled by banks or governments.

3. **Prioritize Privacy**: Consider privacy-focused cryptocurrencies such as Monero (XMR) or Zcash (ZEC) to enhance anonymity and reduce traceability.

4. **Stay Informed**: Monitor global and local regulations so you can adapt before mandates impact your holdings or access.

5. **Diversify Across Assets & Platforms**: Avoid putting all your funds in one place. Use multiple tools and platforms to secure your finance and reduce risk of seizure.

The Hidden Threats of Programmable CBDCs

Organizations like the International Monetary Fund (IMF) and Bank for International Settlements (BIS) have outlined programmable features for CBDCs that include:

- **Restricted Purchases** – Funds that can only be used for approved items.
- **Negative Interest Rates** – Charges applied to encourage spending.
- **Expiring Money** – Funds that disappear after a deadline to stimulate economies.

Real-World Example: Canada, 2022

Again, during the Canada convoy protests, the government invoked the Emergencies Act to freeze bank accounts associated with protesters, without court orders. It is believed that Deputy Prime Minister Chrystia Freeland confirmed the following:

> *"Their trucks can be seized, their corporate accounts frozen, and vehicle insurance suspended."*

Read the source on Wikipedia

This shows how centralized financial control can be activated quickly in times of unrest and why decentralized options matter. Understanding the power and risks of CBDCs is essential. By taking proactive steps through AI-powered DeFi tools, self-custody, and privacy protection, you can maintain your financial autonomy even as the world shifts rapidly toward digital currency systems.

Financial Privacy with AI-Powered Tools

As world governments push for cashless economies, they are also increasing financial surveillance. AI-driven privacy tools help keep your transactions private and secure.

Financial Privacy Under Major Threat

1. Financial institutions and agencies now track transactions, with traditional banking systems, giving full visibility into your spending habits.

2. CBDCs have been discussed as a tool for financial oversight, however there are concerns over personal privacy, blockage of purchases and even impose negative interest rates on your money.

3. AI-powered surveillance systems - Many banks already use AI algorithms to monitor and flag transactions, sometimes leading to frozen accounts for "suspicious" activity.

AI-Powered Privacy Solutions

1. **AI-powered Privacy Coins** – Cryptocurrencies like **Monero (XMR)** and **Zcash (ZEC)** use AI-driven obfuscation techniques to hide transaction details.

2. **AI-enhanced DeFi Platforms** – Some DeFi tools use AI-powered **zk-SNARKs** (zero-knowledge proofs) to enable anonymous transactions.

3. **Decentralized Web Browsers & VPNs** – AI-driven browsers like **Brave AI** and privacy-focused VPNs like:

4. **Proton VPN** AI keep financial transactions secure and untraceable.

5. **AI-generated Anonymous Wallets** – AI tools create disposable wallets that self-destruct after transactions, preventing tracking.

Why this matters You keep financial control and avoid becoming a target of surveillance.

The Future – AI-Powered Financial Freedom & Privacy Tools

Always remember that with convenience, and use of many AI tools, you give AI some level of access. The more helpful AI becomes; you must understand what information you are feeding it. However, with AI-driven DeFi, privacy tools, and alternative financial systems, individuals:

- Avoid banking restrictions
- Keep transactions private
- Invest and trade without government interference
- Go fully digital without losing financial freedom.

The world is shifting fast! How will you adapt, and will you control your wealth?

PART 6: The Rise of CBDCs: Control & Privacy Risks

Chapter 20: Central Authorities & Policy Shifts

AI is rapidly transforming industries, and with new government policies, shifting global economic trends, and the unlimited reach of the internet, now is the best time to leverage AI for business and income generation. Here's why:

AI development and regulation are at the forefront of U.S. government:

- **Funding & Grants for AI Startups**: The government is investing in AI innovation, cybersecurity, and automation, opening doors for entrepreneurs to access funding, tax breaks, and incentives.

- **Regulatory Frameworks**: AI regulations are evolving to protect consumers while encouraging AI-driven businesses. Those who adapt early will have a competitive advantage.

- **AI in Workforce & Economy**: AI is expected to increasingly replace certain jobs while hopefully creating new high-paying opportunities in AI-driven industries. Now is the time to build AI businesses before the market becomes saturated.

Lowering Business Costs & Automating Work with AI

AI tools now allow small businesses and individuals to scale faster without huge upfront investments.

- **Content Creation**: AI automates video production, blogging, and social media marketing—reducing content costs.
- **Customer Support**: AI chatbots handle customer service 24/7, reducing overhead expenses.
- **E-commerce & AI Automation**: AI-powered drop shipping, print-on-demand, and marketing automation make it easier than ever to start an online store.

Example: AI-powered Shopify stores use chatbots, automated product descriptions, and AI-generated ads to drive sales.

Global Internet Expansion & AI Democratization

The internet has removed barriers, allowing anyone to build an AI-driven business that scales globally.

1. Freelancers & Solopreneurs Can Compete with Large Companies using AI-driven marketing, automation, and smart analytics.
2. Remote AI Work Opportunities Are Booming as companies seek AI talent worldwide.

3. AI Monetization Platforms (YouTube, TikTok, OpenAI, GPT marketplaces, etc.) let creators make passive income.

Example: AI-driven YouTube channels generate revenue without creators recording a single video, using AI video generators.

AI Demand is Surging Across Every Industry

Businesses worldwide are rushing to adopt AI. If you offer AI services, build AI-driven businesses, or leverage AI for automation, you'll ride the wave instead of being left behind.

High-Growth AI Business Sectors:

1. **AI Video & Content Generation** – Automating YouTube, blogs, & social media
2. **AI E-commerce** – Print-on-demand, AI marketing, automated drop-shipping
3. **AI Chatbot & SaaS Businesses** – AI-powered customer service & tools
4. **AI-Powered Consulting** – Helping businesses adopt AI efficiently
5. **AI Freelancing & Digital Services** – AI-driven ad creation, copywriting, SEO.

Early Movers Will Dominate the Market

The AI revolution is like the early days of the internet; those who start now will have market share before it becomes oversaturated.

- AI startups are being funded at record-breaking levels
- AI job demand is outpacing traditional tech jobs
- AI-generated businesses are scaling faster than human run businesses

Early adopters will have an edge, while latecomers will struggle to keep up.

Summary: The Age of AI has begun - Now is the Time to Take Control

With AI fueling access to automation, content creation, global business opportunities and an important part of many industries, individuals who explore AI today can take control of how they adapt, grow and succeed in the decade ahead. Take next best steps to leverage AI for income & growth. Here are some examples:

- Start an **AI-driven side hustle** (AI video creation, chatbots, AI marketing)
- Use AI to scale your existing business (automate workflows, increase efficiency)
- Learn **AI monetization strategies** (passive income from AI-generated content)
- Build **AI-driven SaaS or consulting services** to help businesses integrate AI.

PART 6: The Rise of CBDCs: Control & Privacy Risks

The opportunity is NOW, before the AI market becomes overcrowded. Are you ready to leverage it?

AI Ethics & Legal Considerations

AI +Ethics and Responsible Content Creation for Monetization

Transparency & Disclosure

1. **AI Disclosure in Content** – Many countries, including the EU (AI Act) and U.S. states like California, are introducing regulations requiring AI-generated content to be clearly labeled. Transparency helps prevent deceptive practices.

2. **Ethical Sponsorships** – The FTC mandates clear disclosure of sponsored content. AI-generated influencer videos must comply with these regulations to ensure consumers are not misled.

3. **Authenticity Verification** – Integrity digital watermarking and blockchain verification are becoming regulatory standards for AI-generated media. Platforms like YouTube and TikTok may require verification methods to prevent misinformation.

4. **Avoiding Deepfake** – Many more governments, including the U.K. and China, have strict laws against malicious deepfakes. Using AI-generated videos for fraudulent impersonation can lead to criminal liability.

5. **Data Privacy Protection** – AI-generated videos that collect or analyze user data must comply with privacy laws such as GDPR (EU), CCPA (California), and upcoming U.S. federal AI regulations.

Fairness & Bias Prevention

1. **Unbiased AI Training** – Discriminatory AI-generated videos may violate anti-discrimination laws, such as the

2. **U.S. Civil Rights Act and the EU's Digital Services Act** – Companies must ensure AI does not reinforce harmful biases.

3. **Fact-Checking AI Narratives** – AI-generated misinformation could lead to liability under defamation or false advertising laws. Countries like Australia are pushing for stricter penalties against AI-spread disinformation.

4. **AI for Inclusive Representation** – Ethical AI video generation should align with diversity and inclusion policies to promote fair representation and prevent potential regulatory action for discriminatory practices in advertising and media.

5. **Clickbait & Misinformation** – The Digital Services Act (EU) and various FTC guidelines in the U.S. are cracking down on misleading AI-generated content. Platforms must ensure monetized AI videos do not deceive users.

6. **Cultural Sensitivity in AI Creations** – Certain countries have censorship laws regarding offensive AI content. For instance, China strictly regulates content that misrepresents cultural heritage.

Legal & Copyright Compliance

- **Respect Copyrighted Materials** – AI-generated content that uses copyrighted assets without permission can lead to legal consequences under the DMCA (U.S.) and the EU Copyright Directive.

- **Use Open-Source & Licensed Content** – Many AI generative tools operate under strict licensing agreements. Violating these terms may lead to lawsuits or bans from platforms.

- **AI-Generated Voice & Likeness Consent** – Laws like California's Right of Publicity prohibit using someone's likeness without consent. Unauthorized AI voice cloning could lead to legal claims.

- **Comply with AI Content Regulations** – Countries are drafting AI-specific regulations (such as the EU AI Act and the U.S.

Algorithmic Accountability Act) that require AI content creators to follow transparency and accountability guidelines.

- **Ethical AI Licensing Models** – As AI tools become more regulated, licensing agreements will determine what aspects of AI-generated content can be legally monetized without violating terms of service.

Responsible Regulatory Methods

1. **Sustainable AI Video Production** – AI-generated videos that require excessive computational power may fall under scrutiny for environmental impact. The EU is considering carbon footprint regulations for AI systems.

2. **AI Accessibility Features** – Laws like the Americans with Disabilitics Act (ADA) require content to be accessible to all. Monetized AI videos should include captions, translations, and other accessibility features to comply with regulations.

3. **Promote Educational & Beneficial Content** – Regulatory bodies, such as the FCC in the U.S., promote responsible AI content for education while restricting AI-generated videos that spread harmful misinformation.

4. **Avoid Manipulativc Algorithms** – AI monetization strategies must comply with

fair advertising practices. The EU and the U.S. are drafting regulations to curb manipulative AI-generated marketing tactics.

5. **Ethical AI-Generated Ads** – False advertising laws apply to AI-generated marketing content. AI-generated product reviews, testimonials, or exaggerated claims can lead to legal actions.

Chapter 21: AI in Personal Growth & Education

AI's impact on education, personal growth, and the potential closure of public and private schools in the USA is profound and multifaceted. While AI offers opportunities to transform learning and development, it also brings challenges and consequences that could reshape social and psychological dynamics for both the new generation (Gen Alpha and beyond) and existing generations.

1. Impact of AI on Education

AI enables personalized education experiences, where learning paths can be tailored to individual students based on their strengths, weaknesses, and learning styles. This could help close achievement gaps and ensure that every student learns at their own pace. For example:

- AI-powered Platforms like personalized tutoring systems, adaptive learning apps, and intelligent content delivery systems can make learning more accessible, engaging, and effective for students of all ages.

- Predictive Analytics could help teachers identify struggling students early and provide targeted interventions, thus improving overall student outcomes.

PART 6: The Rise of CBDCs: Control & Privacy Risks

Blended and Hybrid Learning Models:

AI can contribute to a shift toward blended or hybrid learning, where traditional in-person education and digital tools work together. This can include:

- AI Tutors: Virtual assistants and AI tutors can provide 24/7 support to students, helping them with homework, understanding complex concepts, and practicing skills outside of class.

- Online Learning Platforms: As AI becomes more integrated into these platforms, they will offer more advanced interactive learning environments, improving accessibility for students in rural or underserved areas.

Teacher Support and Automation:

If not already, AI will automate administrative tasks like grading and content creation, allowing teachers to focus more on individualized instruction and emotional support for students. While this reduces administrative burdens, it may also lead to job displacement in educational roles related to manual grading and content delivery.

Potential School Closures:

If public and private schools close or move toward digital learning platforms, AI will play an integral role in maintaining education continuity. However, this shift raises several concerning challenges:

- **Equity Issues:** Access to AI-driven educational tools requires reliable internet, devices, and technical support. Without equitable access to technology, the digital divide could worsen.

- **Teacher Training:** Teachers will need specialized training to work alongside AI tools effectively. A lack of professional development could hinder the adoption of AI, especially in underfunded schools.

2. AI's Impact on Personal Growth

AI's influence on personal growth will vary depending on how education is delivered and how individuals engage with technology.

Skills Development:

- **Critical Thinking and Problem-Solving:** AI tools can enable students to develop critical thinking and problem-solving skills in a more interactive way. They may be able to simulate real-world scenarios, experiment with various solutions and receive feedback in real-time.

- **Soft Skills:** While AI can enhance technical learning, there's a concern that it might not address emotional intelligence, collaboration, and interpersonal communication as effectively as traditional school settings.

These are essential for personal growth and success in life.

- **Lifelong Learning:** AI-powered platforms will make learning more accessible for people of all ages. Adults could continuously improve skills in various fields, supporting lifelong learning and personal growth opportunities.

AI as a Guide for Self-Improvement:

- AI systems like fitness trackers, mental health apps, and personalized coaching programs could promote self-improvement and well-being, guiding individuals in areas like productivity, emotional intelligence, and mental health.

3. Social and Psychological Impact New Generations (Gen Alpha and Beyond)

Social Isolation and Communication Skills:

Reduced Face-to-Face Interaction: With AI-powered online learning, children and adolescents may have fewer opportunities for in-person socialization. This could impair their ability to develop crucial communication skills, empathy, and conflict resolution.

- **Screen Dependency:** The increased use of screens and AI-driven platforms for learning may lead to a decline in physical healthy activity and social interaction, potentially contributing to isolation, loneliness, and mental health challenges.

- **Altered Social Norms:** As more people, especially younger generations, interact with AI-driven educational tools and digital environments, social norms may shift, with younger generations growing up more comfortable in digital spaces than physical ones.

Identity and Self-Worth:

- **AI-Generated Content and Validation:** Young people growing up in a world with AI-generated content may face challenges related to self-esteem, if they compare themselves to idealized, AI-enhanced portrayals of reality.

- **Perception of Reality:** The rise of AI in video production and social media could make it harder for younger generations to distinguish between authentic and manipulated content, leading to issues around body image, unrealistic expectations, and self-worth.

Psychological Development:

- Emotional Development: AI systems cannot fully replicate human interactions that are critical for emotional development. The lack of face-to-face interactions with peers and adults might stunt the development of emotional intelligence and resilience.

- Increased Pressure for Perfection: AI's ability to track and measure student progress in real-time might increase pressure on students to constantly perform at elevated levels, potentially leading to anxiety or burnout.

4. Impact on Existing Generations (Gen Z and Millennials)

Adapting to Technology:

- **Digital Natives:** Gen Z and Millennials, already familiar with technology, will continue to adapt to AI-enhanced educational tools and work environments. They will leverage AI for self-improvement, business ventures, and educational purposes, potentially speeding up innovation and entrepreneurship.

- **Skill Shifts:** These younger generations may be more likely to upskill themselves with AI tools to stay competitive in the job market. This could lead to a shift in the job market, with greater emphasis on tech-savvy individuals.

Mental Health and Well-Being:

- **AI in Mental Health Support:** Both Gen Z and Millennials could benefit from AI-driven mental health tools. AI-powered apps could provide real-time support for managing stress, anxiety, and depression, especially in a world where traditional support systems (like in-person therapy) might be harder to access.

- **Work-Life Balance:** With more reliance on AI in the workplace, Gen Z and Millennials could benefit from better work-life balance. If needed, AI could handle repetitive tasks, allowing more time for creativity and human interaction.

Speculative Future

As the potential closure of schools increases and the dependency towards AI-driven education becomes greater, the world may experience significant long-term effects:

- **Global Educational Disparities:** While some countries or regions will adapt quickly and efficiently to AI-powered education, others may face substantial challenges due to economic disparities and unequal access to technology. This could exacerbate global educational inequalities.

- **New Roles for Teachers and Educators:** While traditional roles for teachers may diminish in certain contexts, new roles could emerge as educators become facilitators, guiding students through AI-powered learning systems and ensuring emotional and social development. For many seasoned teachers who have had years of experience in traditional teaching roles without using AI, the learning gap may be challenging or even life changing unless there is flexibility and a heroic effort to acquire new learning.

Decentralized Learning Models: Learning could become decentralized, with students taking charge of their own education. Is this possible? With AI, they could design personalized curriculum, choosing topics, resources, and delivery methods that best suit their learning styles.

Summary

AI's impact on education and personal growth will bring both opportunities and challenges. While it offers innovative ways to learn, improve, and grow, it also raises concerns about social isolation, mental health, and the changing role of teachers.

The next generation (Gen Alpha) will likely experience a hybrid of AI-driven education and human socialization, which could drastically alter how they grow socially and psychologically.

Meanwhile, older generations may see shifts in their professional lives and how they engage with technology, pushing them to adapt to a more automated, digital future. As AI transforms education, balancing technological progress with human-centered values will be key to ensuring a positive and inclusive outcome for all.

PART 7

AI, Crypto & The Future of Financial Systems

PART 7: AI, Crypto & The Future of Financial Systems

Chapter 22: Crypto & Control – Your Money, Your Rules

This chapter explores how cryptocurrencies and DeFi can give individuals more control over their money without relying on banks or government systems. Powered by blockchain technology, these systems allow people to send, receive, save, and invest money directly, cutting out the middlemen.

DeFi opens the door to more privacy, freedom, and access, especially for those in underserved regions, but it's important to understand the risks and regulations with DeFi.

Key Takeaways:

What is DeFi? A system that lets you borrow, lend, trade, or invest money without banks or central control.

How Crypto Gives You Control: Cryptocurrencies like Bitcoin or Ethereum let you **own and manage your money directly**, not through a bank or government.

Pros & Cons of Crypto Investment:

Pros: Freedom, high returns, borderless access.

Cons: Volatility, scams, and unclear laws in some countries.

Chapter 23: The Future of AI and Banking

AI is reshaping industries worldwide, and finance is no exception. From algorithmic trading to fraud detection, AI is driving efficiency and innovation. However, alongside these advancements, governments and financial institutions are exploring digital currencies, prompting discussions about privacy, financial control, and economic stability.

Here I aim to provide a balanced perspective on these changes, using what I believe are credible sources to analyze the implications and opportunities for individuals and businesses.

Central Bank Digital Currencies (CBDCs): A Global Perspective

CBDCs have been a topic of discussion among global financial institutions, including the International Monetary Fund (IMF) and the Bank for International Settlements (BIS). The Federal Reserve has also released reports exploring the potential benefits and risks of a U.S. CBDC.

According to the IMF, CBDCs could enhance financial inclusion and provide a secure alternative to cash (IMF Report, 2021). The BIS argues that CBDCs may improve payment efficiency and reduce illicit activities (BIS Working Paper No.976, 2021). However, concerns about privacy and financial oversight have also been raised. The Federal Reserve's 2022 discussion paper on CBDCs highlights potential risks, including cybersecurity threats and the impact on commercial banking.

PART 6: The Rise of CBDCs: Control & Privacy Risks

Consider:

- CBDCs could provide faster, more secure transactions, a reduction in fraud, and improvements in payment efficiency, but may also increase government oversight of financial activity.

AI plays a role in financial surveillance and risk management. AI-powered transaction monitoring systems are already helping banks detect fraud and money laundering (Federal Reserve Bank of New York, 2022). While beneficial, these systems raise questions about financial privacy, particularly when coupled with CBDCs.

According to the BIS, AI-driven financial systems could allow for greater regulatory oversight, which may help prevent illicit activities but could also raise concerns about personal financial autonomy (BIS Innovation Hub, 2022). Balancing regulatory requirements with financial privacy is a key challenge in the evolving financial landscape.

Privacy – AI Considerations & Risks

Governments have discussed programming CBDCs to limit certain purchases or impose expiration dates on money (BIS Annual Economic Report, 2021). The potential for negative interest rates on CBDCs could incentivize spending but also impact savings behavior (IMF Fintech Note, 2021).

Privacy Considerations:

1. How much transaction data is collected?

2. Who can access each person's behavioral spending?

3. Are there protections in place for mass surveillance? Who has restricted access?

4. Is there human review and how much over such automated enforcement?

5. If CBDCs are implemented, what happens if AI, connectivity, and cloud providers in general fail?

6. Can we see an increase in identity fraud?

7. Will there be targeted attacks against AI-driven fraud systems?

8. Will the needed regulations come before the vast implementations?

Decentralized Finance (DeFi): A Viable Alternative?

DeFi offers an alternative to traditional banking. DeFi applications operate on blockchain networks and allow users to store, invest, and transact without intermediaries. Bitcoin and other cryptocurrencies have been cited as potential safeguards against inflation and financial control measures (IMF Crypto Policy Report, 2023).

DeFi also carries risks:

- **Security Concerns**: Hacks and smart contract vulnerabilities have led to significant losses.
- **Regulatory Uncertainty**: Governments are still formulating policies around DeFi (BIS Crypto Report, 2023).
- **Volatility**: Crypto markets remain highly speculative (IMF Financial Stability Report, 2022)

Additional Steps to Protect Your Wealth in the Age of AI

Given these developments, individuals can take proactive steps to safeguard their financial future:

1. **Diversify Holdings**: A mix of traditional assets (gold, stocks) and digital assets (Bitcoin, DeFi) can provide financial resilience (IMF Economic Outlook, 2022).
2. **Use Privacy-Preserving Technologies**: AI-driven privacy tools, such as zero-knowledge proofs, enhance financial anonymity (BIS Fintech Paper, 2022).
3. **Stay Informed**: Understanding regulatory changes and financial trends is crucial for making informed decisions (Federal Reserve Policy Reports, 2023).

Summary: The Path Forward

AI, CBDCs, and DeFi are reshaping the financial world.

While CBDCs could enhance efficiency, they also raise questions about oversight and personal financial freedom. DeFi presents an alternative but comes with its own risks. The key is to approach these developments with a well-informed, balanced perspective, leveraging AI and digital finance to empower rather than restrict individuals.

The next 10 years will bring huge changes in banking and finance, and AI will continue to play a key role in how money is handled.

Key Takeaways:

- AI in Finance & Content Creation – AI will make financial decisions, create personalized banking services, and generate content with minimal human effort.

Dangers of AI-Driven Finance – AI could increase financial control by governments and banks, limiting how people access and use their own money.

- How to Stay Ahead – Learning AI skills, investing in digital assets, and using decentralized finance can help you create and maintain financial freedom.

Example: AI could automate financial transactions, but if banks control AI, they might restrict access to funds so being prepared with alternative financial tools is key.

PART 6: The Rise of CBDCs: Control & Privacy Risks

Sources:

- Satoshi Nakamoto's Bitcoin Whitepaper – The foundation of cryptocurrency (bitcoin.org).
- Ethereum.org – Information on smart contracts and DeFi (ethereum.org).
- U.S. Federal Reserve Reports on Digital Currencies – Insights into government oversight of crypto (federalreserve.gov).

Books on AI & Finance:

- Life After Google by George Gilder (Explores decentralization and the future of finance).
- The Age of Cryptocurrency by Paul Vigna & Michael Casey (Explores how Bitcoin is reshaping money).
- AI Superpowers by Kai-Fu Lee (The future of AI-driven economies).

Key Points: Is AI Seen as a Threat?

AI is seen by many as controversial or a challenge. However, if framed as a guide to financial empowerment for human survival AI can be seen as insightful and valuable. If framed as an attack on banks, people and governments, AI can be seen as controversial or a real challenge. The goal should be to first educate everyone so that each person can feel confident in taking the necessary actions, to provide solutions for protection of wealth, to invest smartly, and adapt to the changing financial AI landscape.

By staying informed and proactive, I hope individuals can navigate this changing landscape and make decisions that align with their financial goals and values.

The future belongs to those who are prepared.

Will you be one of them?

PART 6: The Rise of CBDCs: Control & Privacy Risks

Chapter 24: Aging – It is Inevitable

Aging is the one journey every single one of us is guaranteed to take, no matter what our race, background, gender, success, or ambition. It is a shared human experience, and yet, it is often ignored in conversations about innovation and the future. But as the world turns faster with technology and AI, we must ask ourselves: ***What kind of future are we building if we leave behind the very people who raised us, taught us, and shaped our values?***

We will all arrive at that stage of life when our pace slows, our strength changes, and our needs evolve. When the world feels different, not just because of age but because the world is changing around us. Some of us are already there. Others are watching beloved parents or grandparents try to make sense of a digital world that barely pauses long enough for them to grasp.

Aging Should Never Mean Becoming Invisible

- What if the future didn't leave anyone behind?
- What if technology could become a gentle companion, not an overwhelming challenge?
- What if financial tools were simple, safe, and powerful designed with dignity in mind?

YOUR MONEY IN THE AGE OF AI

This chapter is for those who may feel like the world is moving too fast. It's also for the younger generations who want to ensure their loved ones aren't left behind. Because as we empower the elderly, and show respect, we also help prepare for *our own future selves*.

AI, Digital finance, and Intentional Innovation.

Let's explore how AI, digital finance, and intentional innovation can help the aging population not just to *survive*, but to conquer and survive with grace, confidence, and power.

The Reality

- Many older adults are living longer, but with **fixed incomes**, increasing medical expenses, and shrinking retirement savings.

- The **digital divide** hits hard here: they're often the last to benefit from tech, AI, or digital financial tools.

The Challenge

- Tech platforms are rarely designed with elderly usability in mind.

- Fears of fraud, scams, or complexity keeps many away.

- Social stigma: Some think, "It's too late for me to learn."

The Shift is Happening

- Grandparents use Zoom and FaceTime.

- Seniors are banking online, managing health records, and joining social platforms.
- AI tools like **voice assistants**, **simplified interfaces**, and **learning apps** are making tech more accessible.

Realistic, Full-Proof Tools

Solutions	Description
• Voice-activated tech (like Alexa, Google Assistant)	No typing needed, perfect for those with vision or motor issues. Can be used to manage appointments, set reminders, or even make payments.
• Simplified AI financial tools	Platforms like Cleo, You Need a Budget (YNAB), or robo-advisors with human support options help older users invest or budget safely.
• Banking apps with "senior mode" or caregiver access	Some banks offer simplified views or allow trusted caregivers to monitor without control.
• Online learning designed for older adults	Tools like GetSetUp, Senior Planet, and TechBoomers provide tutorials.
• Tablet-based solutions with larger icons and AI support	Devices like GrandPad are built just for seniors.

• Community workshops and intergenerational learning	Pairing seniors with younger "digital mentors" boosts confidence and inclusion.

Key Takeaway:

The aging population isn't a problem to solve; it's powerhouse of wisdom, ready to thrive in the digital age. AI, if designed and delivered thoughtfully, can help elderly people survive with dignity and independence.

PART 8

AI & Banking Meet Tribal & Rural Communities

Chapter 25: AI & Banking in Tribal and Rural Areas

In today's fast-moving world of artificial intelligence, digital currencies, and smart financial tools, it's easy to forget that millions of people around the world still live without internet access, bank accounts, or even electricity.

From small farming villages in parts of Africa, to mountain communities in South America, tribal regions in Asia, and remote Aboriginal communities in Australia, many cultures have lived for generations using local trade systems, shared resources, and oral traditions to survive. Money, as we know it, isn't the center of their world. Trust and relationships are the foundation of how they live, thrive, and support each other.

The rise of AI in finance isn't just a story for big cities or wealthy nations. It's a global shift. And as we move forward, we are tasked to make sure that progress doesn't come at the cost of dignity, culture, or independence. Financial freedom should be a gift not a takeover; but what happens when governments, charities, or corporations begin introducing digital financial tools, like AI-powered banking apps or government-issued digital currencies, into these untouched areas?

PART 8: AI & Banking Meet Tribal & Rural Communities

The Promise: Access, Opportunity & Empowerment

Some may see AI as disruption and others may see it as progress. Technology can certainly bring hope, when introduced responsibly.

AI and decentralized finance (DeFi) could possibly:

- Help farmers access fair microloans or crop insurance using AI-powered platforms.
- Allow families to receive international donations without banks taking large fees.
- Teach young people how to build wealth using AI-guided tools, no formal banking is required.
- Provide financial identity to those who've been historically left out of the system.

In places where banks are miles away or where inflation has destroyed the value of local money, AI and digital tools could give people more control, more choices, and more power to grow their communities.

The Challenge: Control, Culture & Consequences

But there's another side we must consider, especially when financial technology is introduced without proper education, consent, or understanding.

1. What if people are pushed into using government-run digital currencies that track every transaction?

2. What if communities begin to lose traditional ways of trade, sharing, and support because everything becomes digitized?

3. What happens when a village no longer uses livestock or goods as trade but now needs smartphones and internet access just to survive?

For some tribal or rural areas, the sudden arrival of digital finance may feel more like control than freedom. The shift could threaten cultural traditions, widen inequality, or create a new kind of dependency where the power still belongs to outsiders.

Why Digital Finance Should Matter to You

You may not live in a village without banks, but the questions being raised are still universal:

1. Who decides how people use money?

2. Should financial freedom mean the same thing for everyone?

3. How can AI and technology serve people without taking away that which makes their communities unique?

These are conversations that matter, not just for tech experts or economists but for students, parents, entrepreneurs, and policymakers alike. Because how we bring technology to those who have lived without it may shape the future of freedom for everyone.

What We Can Do Moving Forward

1. Support tech initiatives that work with communities, not **over** them.

2. Educate ourselves and others about how AI and DeFi can be empowered when introduced with wisdom and care.

Ask better questions, not just about what's possible, but about what's right.

Chapter 26: Digital Finance for the Marginalized

Across Australia, Africa, Asia, and Latin America, a quiet revolution is already happening AI, mobile banking, and DeFi are beginning to reach communities once excluded from the global financial system. When done with care and cultural understanding, these tools can bring real transformation.

Indigenous Australia: Digital Tools for Cultural and Financial Resilience

In remote Aboriginal communities across Australia, financial exclusion has long been a challenge. Many people live hours away from the nearest bank, with limited access to mainstream financial services. Yet these communities possess rich cultural and economic systems built on reciprocity, kinship, and collective stewardship of the land.

In recent years, initiatives have emerged that aim to bridge the gap between traditional practices and modern digital finance while respecting the unique identity of Indigenous Australians.

Digital Banking for Remote Communities

The Australian government and not-for-profits like First Nations Foundation have launched programs that:

- Deliver mobile banking units to communities where there are no bank branches.

- Provide digital literacy training for elders and youth alike.
- Help Indigenous Australians recover lost superannuation (retirement) funds via digital tools.

Preserving Culture Through Tech

Some communities are also using digital platforms to:

1. Record and protect Indigenous languages and stories.
2. Sell art and crafts directly through online marketplaces, reducing exploitation.
3. Explore blockchain to create Indigenous-owned systems of land rights registration and cultural asset tracking.

Why Culture Through Tech Matters:

Introducing financial tech in these communities isn't just about access, it's about empowerment, cultural preservation, success, and ownership. Done right, these tools allow Indigenous Australians to participate in the digital economy without surrendering their values or sovereignty.

Mobile Money in Africa

Take the case of **M-Pesa** in Kenya and other parts of Africa. M-Pesa, a mobile money platform, allows users to send, receive, and save money using simple, feature phones no bank account or smartphone needed. Rural families can now receive school payments or emergency funds from relatives working in cities. Small shop owners can accept mobile payments safely without cash on hand. Women in remote areas now have their own financial identity and autonomy.

Why Mobile Money in Rural Areas Matter

M-Pesa shows how technology adapted to local needs can empower people economically without erasing their culture or requiring expensive infrastructure.

AI in Farming: Helping Farmers Make Smarter Decisions

Organizations like Hello Tractor (Nigeria) and Digital Green (India) are using AI to help rural farmers:

1. Track crop yields and soil conditions for better productivity.

2. Access shared equipment through smart contracts and GPS tracking.

3. Receive tailored agricultural advice via SMS in local languages.

Why AI in Farming Matters

By connecting traditional farming with AI-driven tools, these communities can improve food security, earn better incomes, and stay independent.

PART 8: AI & Banking Meet Tribal & Rural Communities

Blockchain for Transparency & Microfinance

Platforms like Kiva Protocol use blockchain technology to create secure digital identities and track microloans in underbanked areas. Citizens in Sierra Leone are using Kiva to access loans without credit history or traditional IDs. The transparency of blockchain helps prevent fraud and ensures donations reach the intended recipients. DeFi can help bypass corrupt systems, reduce dependency on central banks, and give more people a seat at the financial table.

Chapter 27: Balancing Traditions with Technology

When introduced thoughtfully, AI and digital tools can serve as bridges, not bulldozers. It's crucial to remember: Financial empowerment doesn't mean replacing ancient ways of life. It means giving people options, not taking away freedom. The most successful programs:

1. Respect local customs and decision-making.
2. Offer training and digital literacy.
3. Let the community lead the adoption process.

A Shared Future

From the streets of New York to the villages and tribes of remote areas in the world, one truth remains clear: financial awareness is not just a tech issue, it's a human issue. Let's build AI systems that serve all people, regardless of zip code, background, or connection speed. Whether you're a young adult navigating your first paycheck or a retired person watching the world change, this moment belongs to all of us.

It's Your Money in the Age of AI Boost Your Income & Take Control Today

Epilogue: Facing the Future

Thank you for reading my book. More importantly, thank you for choosing to think intentionally about how to use AI in your life. **That choice matters more than it may seem.** As we all stand at the crossroads of artificial intelligence, digital finance, and global transformation, one truth becomes clear: **The future belongs to those who are willing to adapt, learn, and act, because AI is no longer something distant or optional. AI is already embedded in our everyday use.**

Throughout this book, you've discovered the incredible impact AI is having in every aspect of our lives, in finance, and in the world, from intelligent wealth-building strategies to AI video monetization, decentralized finance, and the evolution of digital currencies. You've explored real-world examples, practical tools, privacy, and powerful opportunities that are reshaping how we create, earn, and grow. But as you know, knowledge is power, and it is the first step.

- It's time to take control. Now is the time to leverage AI tools for smarter, faster financial decisions.
- Unlock new income streams using AI-powered content and digital platforms.
- Stay informed and flexible as modern technologies reshape our global economy.
- Ask others to read this book.

YOUR MONEY IN THE AGE OF AI

The future of wealth-building is no longer a distant idea, **so, take control of your financial future, don't wait for change to find you; be the one who shapes it.**

Let this not be the end but the spark that launches your journey into a more empowered, financially free future. AI is continuously unfolding right before our eyes and by embracing many of the strategies and mindsets explored in this book, you can position yourself ahead of the curve.

There are all sorts of AI tools out there, do your research, and as AI continues to evolve, remember what remains under your control, it is how you choose to engage with AI. The question isn't whether the world will change.

Will you be ready to lead in this change?

The tools are in your hands. The moment is now!

Acknowledgements

This book is dedicated to those who crave awareness, discernment, and financial freedom in these changing times; along with those who are suffering the challenges of unemployment. This book is also dedicated to all my family members who I dearly love; to whom I wish happiness, health and peace.

I pray that this book will also help everyone, not only for financial freedom but also to uplift your spirit to see the world in a different light.

I acknowledge and give thanks to God almighty, Father, Son and Holy Spirit from whom all blessings flow, the one who has sustained me through very tough times, who took me out of the valley, who walked with me through the dry seasons to green pastures and who led me to still waters. He alone has sustained me.

God shall be magnified overall.

Appendix: 100+ Monetization Strategies

AI-generated videos can be monetized in several ways. Whether you create content for social media, business clients, or personal branding, here are 100+ revenue streams:

Ad Revenue & Platform Monetization

1. YouTube Partner Program (AdSense)

Example: Tech YouTubers who use AI-powered video editing and analytics to optimize content for YouTube ads, generating significant revenue.

2. Facebook In-Stream Ads

Example: Viral video pages like "UNILAD" use AI tools to auto-edit videos and insert mid-roll ads, increasing ad revenue.

3. TikTok Creator Fund

Example: AI-generated animation accounts on TikTok, like "AI Art Universe," earn from the Creator Fund by posting AI-generated story videos.

4. Monetizing Instagram Reels (Bonuses & Ads)

Example: AI-generated motivational quote videos with voiceovers and animations monetize through Instagram Reels bonuses.

5. LinkedIn Video Ads Revenue

Example: AI-powered marketing firms use LinkedIn AI video ads to generate leads and earn ad revenue through partnerships.

6. Twitter/X Video Monetization (Ads & Subscriptions)

Example: AI-powered news channels monetize breaking news clips using automated video generation and ad-sharing on Twitter/X.

7. Twitch Video Ads & Subscriptions

Example: Reportedly, AI-driven virtual streamers like CodeMiko use deepfake and AI-enhanced animations to attract Twitch subscribers and ad revenue.

8. Snapchat Spotlight Payouts

Example: AI-generated short-form comedy skits have earned thousands through Snapchat Spotlight payouts.

9. Rumble Video Monetization

Example: AI-generated explainer videos on alternative platforms like Rumble earn through ad-sharing programs.

10. Medium Partner Program for AI Video Articles

Example: Writers create blog posts about AI-generated videos and embed them in Medium articles, earning revenue via the Partner Program.

Affiliate Marketing

11. Promote Amazon Affiliate Products in AI-Generated Videos

Example: AI-generated unboxing videos promote Amazon products with affiliate links, leading to commissions on sales.

12. Affiliate Links for Online Courses & Digital Products

Example: AI-generated explainer videos promote Udemy AI courses, generating affiliate revenue.

13. Software Affiliate Programs (Adobe, Canva, etc.)

Example: YouTube tech reviewers use AI to create tutorials on Adobe Premiere Pro and earn affiliate commissions.

14. AI Tool Referral Programs (Synthesia, Pictory, Runway ML)

Example: AI content creators refer viewers to Synthesia for AI video generation and earn per subscription.

15. VPN & Security Software Affiliate Promotions

Example: AI-generated cybersecurity awareness videos promote VPN affiliate programs.

16. Stock Footage & Music Affiliate Sales

Example: AI-generated B-roll footage channels on YouTube promote Envato affiliate links for stock

17. Travel Booking & Accommodation

Example: AI-generated travel videos featuring hotel

recommendations are monetized via Booking.com affiliates.

18. Gadget & Tech Product Affiliate Marketing

Example: AI-generated animated gadget review videos drive sales via Best Buy and Amazon affiliate programs.

19. Hosting & Domain Affiliate Programs

Example: AI-generated tutorial videos recommend Bluehost and SiteGround hosting affiliate links.

20. AI Art and Design Tool Referrals

Example: AI-generated time-lapse art videos promote Midjourney and DALL·E with affiliate links.

Sponsored Content & Brand Deals

21. Paid Sponsorships from AI Software Companies

Example: AI video creators partner with Runway ML to showcase AI-powered editing features.

22. Collaborate with SaaS Brands for Video Promotions

Example: AI-generated SaaS explainer videos help tech startups promote software tools.

23. Develop an AI Video SaaS Product

Example: Develop a software-as-a-service (SaaS) platform that allows businesses or individuals to create AI-generated videos themselves.

1. **Freemium Model:** Offer free basic services (limited video generation, templates, etc.) and charge for premium features like

custom templates, additional content length, or higher quality video rendering.

2. **API Integrations:** If you have a scalable product, you can provide an API for businesses to integrate AI video creation into their apps, websites, or services.

24. Influencer Marketing Campaigns

Example: AI-generated avatar influencers work with Nike for digital advertising campaigns.

25. Paid Video Testimonials & Endorsements

Example: AI-generated avatars provide personalized brand endorsements for E-commerce stores.

26. Sponsored Unboxing & Product Review Videos

Example: AI-generated voiceovers narrate sponsored product reviews for tech gadgets.

27. Brand Collaborations for Explainer Videos

Example: AI agencies create product explainer videos for FinTech startups. Also, A freelancer selling AI marketing videos on Fiverr can make substantial income per month by creating just a few videos and receiving sponsorships per week.

28. Creating Ad Content for Companies

Example: AI-generated marketing videos are sold to Shopify store owners.

29. Video Placement Deals with Brands

Example: AI-driven animation channels insert sponsored brand integrations.

30. Native Advertising Integrations in Videos

Example: AI-generated educational videos subtly promote sponsor brands.

31. Sponsoring AI Video Software Tutorials

Example: AI tutorial YouTubers receive sponsorship from AI video tool companies.

Course Creation & Educational Content

32. Sell AI Video-Making Courses on Udemy

Example: AI instructors teach Synthesia and Pictory automation on Udemy.

33. Create AI Content Creation Training on Teachable

Example: AI coaches launch premium AI editing courses.

34. Offer AI Video Production Workshops

Example: AI-powered webinars train businesses on AI content automation.

35. Monetize Skillshare AI Video Editing Courses

Example: AI-generated Skillshare courses on AI filmmaking earn passive income.

36. Sell Private AI Video Coaching Sessions

Example: AI video editors offer personalized coaching via Zoom.

37. Write & Sell eBooks About AI Video Creation

Example: AI entrepreneurs publish eBooks about automating content with AI.

38. AI Training for Corporate Teams

Example: AI video trainers teach AI-powered marketing to businesses.

39. Paid Memberships for Exclusive AI Video Content

Example: Patreon creators offer AI video tutorials to subscribers.

40. Online Bootcamps for AI Video Mastery

Example: AI video bootcamps provide in-depth training on video automation.

41. White-Label AI Video Training for Companies

Example: Businesses purchase AI-powered video training courses for employees.

Content Licensing & Digital Asset Sales

42. Sell AI-Generated Stock Videos on Shutterstock Adobe Stock

Example: AI artists upload AI-generated aerial footage to stock sites and earn royalties.

43. Create & Sell AI Video Templates (Canva, Envato, etc.)

Example: AI video creators design customizable video templates for social media marketing.

44. License AI-Generated Explainer Videos to Companies

Example: Startups buy AI-created whiteboard animation videos for corporate training.

45. Sell AI-Generated Animated Clips for Commercial Use

Example: AI-generated cartoon characters are licensed to YouTube animation channels.

46. Offer AI Voiceover & Narration Services

Example: AI-generated voiceovers are sold to podcasters and audiobook creators.

Example: AI-narrated educational videos using tools like ElevenLabs for voiceovers and Midjourney for visuals. Ad revenue and sponsorships help earn additional cash.

47. Sell AI-Powered Video Intros & Outros

Example: AI-designed YouTube intros and outros are sold on Fiverr. Freelancers can provide high-quality content faster than traditional methods, for new clients and digital services.

48. License AI-Generated Backgrounds & Green Screen Footage

Example: AI-generated virtual backgrounds are sold to video editors.

49. Sell AI-Enhanced Time-Lapse Videos

Example: AI-generated hyper-lapse cityscape videos are sold as stock footage.

50. Create & Sell AI-Generated Text-to-Video NFT Art

Example: AI video artists mint and sell AI-created NFT animations.

51. License AI-Generated Sports Highlights for Small Media Outlets

Example: AI-generated sports recap videos are sold to independent blogs.

Freelance AI Video Production Services

52. Provide custom video creation for clients:

1. **Subscription-Based Model:** Offer video content generation as a subscription service, with different pricing tiers based on volume or complexity (e.g., monthly plans for YouTube creators, marketing agencies, etc.).

2. **Pay-Per-Video Model:** Charge clients for each video you produce (ideal for businesses needing occasional promotional videos or video explainer).

3. **White-Label Video Services:** Create AI-generated video content and sell it under a client's brand, especially in niches like marketing, e-commerce, and education.

53. Offer AI-Powered Video Editing on Fiverr & Upwork

Example: AI editors provide quick turnaround video edits using Runway ML.

54. Create AI-Generated Business Promo Videos for Clients

Example: AI-generated business intro videos are sold to small brands.

55. Provide AI-Generated Social Media Ads for Companies

Example: AI-generated TikTok ad campaigns are created for E-commerce brands.

56. Offer AI-Generated Whiteboard Animation Services

Example: AI-created whiteboard animations explain concepts for corporate clients.

57. Sell AI-Powered Voiceover Services for YouTube Creators

Example: YouTubers use AI voiceovers to narrate their content.

58. Create AI-Generated Educational Videos for Schools

Example: AI-powered explainers simplify STEM topics for online classrooms.

59. Offer AI-Generated Real Estate Video Tours

- Many businesses need video content but lack the time or budget to produce it. Realtors and AI powered video creation services can benefit with the use of AI-generated property walkthroughs to showcase homes.

Example: A freelancer can offer AI-generated real estate videos, providing agents with virtual property tours for $100 per video. By delivering 20+ videos per month, they may earn $2,000+ monthly.

Example: A marketing agency may use AI to create branded social media ads for small businesses. If they charge $300 per video and have 30+ repeat clients, they potentially can generate approximately $9,000 per month.

60. Generate AI-Powered Personalized Video Messages

Example: AI-generated birthday video messages are sold on Cameo.

61. Provide AI-Generated Meditation & ASMR Videos

Example: AI creates soothing visual and soundscapes for meditation YouTube channels.

62. Sell AI-Generated AI Avatar Spokesperson Videos

Example: AI avatars create spokesperson videos for digital marketing agencies.

Subscription & Membership Models

63. Start an AI Video Subscription Service (e.g., Monthly Video Drops)

Example: AI video creators offer premium explainer videos as part of a monthly membership.

64. Launch an AI Video Patreon Membership for Exclusive Content

Example: AI storytelling channels provide timely access to AI-generated series.

65. Create a Paid Community for AI Video Creators (Discord, Telegram, etc.)

Example: AI-powered animation groups charge a membership fee for industry insights.

66. Offer a Private AI Video Newsletter Subscription (Substack, Ghost, etc.)

Example: AI-generated content trends are shared in a paid newsletter.

67. Sell AI Video Presets & LUTs for Video Editors

Example: AI-created color grading presets are sold for DaVinci Resolve and Premiere Pro.

68. Launch an AI-Powered Video Editing SaaS Subscription

Example: A cloud-based AI video editing tool charges users monthly for automation services.

69. Offer AI Video Style Transfers as a Subscription Service

Example: AI-generated Van Gogh-style video effects are provided as a paid feature.

70. Charge for AI-Powered Sports & Gaming Highlights Clips

Example: AI-powered gaming highlight reels are offered as a premium service for streamers.

71. Create a Subscription-Based AI-Powered Children's Story Video Platform

Example: AI narrates and animates bedtime stories, available via a paid app.

72. Sell AI-Generated News Summaries as a Subscription Model

Example: AI compiles short video news summaries for a premium audience.

Product Creation & E-commerce

73. Sell AI-Generated Animated NFTs as Digital Collectibles

Example: AI-generated looping animations are sold on OpenSea as rare NFT collections.

74. Create & Sell AI-Powered Virtual Reality (VR) Videos

Example: AI-generated VR meditation experiences are sold on Oculus.

75. Offer AI-Powered Video Merchandise (T-Shirts, Posters, etc.)

Example: AI-generated animations are printed on demand and sold as posters.

76. Sell AI-Powered Music Video Lyric Visualizers

Example: AI-generated lyric videos are sold to independent musicians.

77. Sell AI-Generated Commercial Jingles & Music Videos

Example: AI-powered music video visuals are created for brands.

78. Offer AI-powered Birthday & Holiday Video Greetings

Example: AI-personalized holiday greetings are sold on Etsy.

79. Offer AI-Powered Personalized Video Ads for Businesses

Example: AI-customized video ads help small businesses create cost-effective campaigns.

80. Sell AI-Generated Digital Avatars for Virtual Influencers

Example: AI-designed digital influencers are licensed for marketing campaigns.

Emerging AI-Powered Video Business Models

81. Develop AI-Generated Short Films for Streaming Platforms

Example: AI-generated horror films are pitched to Netflix and Amazon Prime.

82. Create an AI-Powered YouTube Kids Channel with Educational Content

Example: AI-generated animated children's songs gain millions of views.

83. Monetize AI-Generated Real-Time Sports Analysis Videos

Example: AI-driven sports commentary videos break down highlights in real-time.

84. Sell AI-Powered Political Commentary & Debate Videos

Example: AI-generated analysis videos on elections gain traction.

85. Offer AI-Generated Wedding & Event Highlight Reels

Example: AI-generated wedding recap videos are sold as a premium package.

86. Monetize AI-Powered Paranormal & Mystery Storytelling Videos

Example: AI-generated mystery narration videos attract millions of views.

87. Create AI-Powered Fact-Checking Video Channels

Example: AI-driven debunking videos provide real-time myth-busting.

88. Sell AI-Powered Video Analysis for Business Intelligence

Example: AI-generated trend analysis videos help investors make decisions.

89. Offer AI-Powered Language Translation & Dubbing Services

Example: AI-powered dubbing tools translate videos for international audiences.

90. Sell AI-Powered Legal & Compliance Training Videos

Example: Law firms use AI videos to train staff in compliance.

91. Develop AI-Generated Health & Fitness Coaching Videos

Example: AI-powered virtual personal trainers create workout routines for YouTube monetization.

92. Create AI-Powered Automotive Review & Tech Breakdown Videos

Example: AI-generated CGI car breakdowns gain ad revenue on tech-focused YouTube channels.

93. Sell AI-Powered 3D Virtual Tour Videos for Real Estate & Tourism

Example: AI-generated VR real estate tours are sold to agents and property developers.

94. Monetize AI-Powered Wildlife & Nature Documentary Clips

Example: AI-enhanced wildlife footage is sold to documentary filmmakers.

95. Develop AI-Powered Localized Video News for Small Towns & Communities

Example: AI-generated hyper-local news reports attract advertising revenue.

96. Offer AI-Generated Custom Animated Music Visuals for DJs & Artists

Example: DJs and electronic artists purchase AI-created visuals for live performances.

97. Create AI-Powered Fashion & Beauty Tutorial Videos

Example: AI-generated virtual makeup tutorials get brand sponsorships.

98. Monetize AI-Generated Historical Reenactment & Storytelling Videos

Example*:* AI-enhanced historical deep-fake storytelling attracts museum and media partnerships.

99. Offer AI-Powered Interactive Choose-Your Own-Adventure Videos

Example: AI-generated interactive storytelling content gets monetized via gaming platforms.

100. YouTube Ad Revenue

Example: Upload AI videos on YouTube and earn through AdSense.

101. Affiliate Marketing Promote products in AI-generated review videos and earn commissions for each sale.

Example: Some online reports suggest that high performance AI-assisted YouTube channels can generate substantial monthly income affiliate commissions; however, these results are not guaranteed and depend on strategy, content quality, and audience engagement from affiliate commissions.

102. Online Courses – Use AI-generated Lectures and Training videos for paid courses.

Example: A graphic designer can use AI avatars and voiceovers to teach logo design, deliver lessons.

Example: A fitness coach could generate AI driven workout demonstration videos. These AI-enabled workflows allow creators to scale course production efficiently, engage students in multiple formats, and reach global audiences.

103. Freelancing – Offer AI video services on platforms like Fiverr and Upwork.

Example: Freelancers can offer AI-generated video creation services on platforms such as Fiverr and Upwork.

Bibliography & References

Amazon Web Services. (n.d.). *Synthesia makes AI video production effortless with generative AI.*

Anderson, K. (2022). *The Future of Career Growth with AI Learning Platforms.*

Baker, P. (2023). *AI and the Writers Guild Strike: The Fight for Creative Control.*

BBC News. (2023). *China Mandates Watermarking for AI-Generated Deepfakes.*

Bender, E. M., et al. (2021). *On the Dangers of Stochastic Parrots: Can Language Models Be Too Big?*

Bis.org. (2023). Central Bank Digital Currencies: A New Era of Control? https://www.bis.org

Biswas, S. (2023). *China's Digital Yuan: The Future or a Surveillance Tool?* BBC News. https://www.bbc.com/news/business-56768580

Brown, S. (2021). *AI-Powered Coaching: How Virtual Mentors Enhance Professional Growth.*

Brynjolfsson, E., & McAfee, A. (2022). *The Business of AI: How Artificial Intelligence Can Transform Companies.*

Buolamwini, J., & Gebru, T. (2018). *Gender Shades: Intersectional Accuracy Disparities in Commercial Gender Classification.*

Carlson, M. (2022). *Automating the News: How Algorithms Are Rewriting the Media.*

AI-Powered YouTube Automation (TechCrunch, 2024)

Casey M and Vignia P. *The Age of Cryptocurrency.* (Explores how Bitcoin is reshaping money).

Chen, J. (2023). What is DeFi? Investopedia. https://www.investopedia.com/defi-decentralized-finance-5113835

Chesney, R., & Citron, D. (2019). *Deepfakes: A Looming Challenge for Privacy, Democracy, and National Security.*

Clark, A., & Leung, C. (2023). *How AI Is Revolutionizing Personal Branding Strategies.*

CNBC. (2023). Bank Closures Continue, Especially in Rural America. https://www.cnbc.com

Dastin, J. (2018). *Amazon Scraps Secret AI Recruiting Tool That Showed Bias Against Women.* Reuters.

Davenport, T., & Ronanki, R. (2018). *Artificial Intelligence for the Real World.* Harvard Business Review.

De Filippi, P., & Hassan, S. (2018). *Blockchain Technology as a Regulatory Technology: From Code is Law to Law is Code.*

First Monday. https://firstmonday.org/article/view/7113/5657

Decentralized Finance Overview – Chen, J. (2023).

"What is DeFi?" *Investopedia*

Bibliography & References

DeepBrain AI - AI Avatar Technology. https://www.deepbrain.io

Digital Green. (n.d.). *Empowering Farmers through Digital Technology*. https://www.digitalgreen.org

Ethereum.org – Information on smart contracts and DeFi (ethereum.org).

European Commission. (2024). *The EU AI Act and Its Implications*.

Federal Reserve Bank. (2023). Trends in Branch Banking. Https://www.federalreserve.gov/publications.htm

FDIC. (2022). Community Bank Study. https://www.fdic.gov/resources/community-banking/report/

Finland's basic income experiment. :Government Publications+3YouTube+3Basic Income Today+3

First Nations Foundation (Australia) – "Indigenous Financial Resilience: Money Stories and the Path Forward" (2020).

https://www.fnf.org.au/research.

Fitzpatrick, K., et al. (2017). *Evaluating the Effectiveness of Woebot, an AI Chatbot for Mental Health Support*.

Ford, Martin. (2015). Rise of the Robots: Technology and the Threat of a Jobless Future. Basic Books, 2015.

Garcia, M., & Lin, P. (2019). *The Effectiveness of AI-Driven Language Learning Apps*.

GDPR. (2022). *General Data Protection Regulation – Data Protection Rules.*

Ghosh, A. (2022). *The Connectivity Divide in Global Development.* Development Technology Reports.

Gomez-Uribe, C. A., & Hunt, N. (2015). *The Netflix Recommender System: Algorithms, Business Value, and Innovation.* ACM Transactions on Management Information Systems.

Goyal, M., et al. (2014). *Meditation Programs for Psychological Stress and Well-Being: A Systematic Review.*

Holmes, W., et al. (2019). *Artificial Intelligence in Schools: Promises and Challenges.*

Hour One. (n.d.). *AI video case studies.* Hour One, Retrieved March 31st, 2026.

Huang, M.-H., & Rust, R. T. (2021). *Artificial Intelligence in Mental Health: Opportunities and Challenges.*

IMF & World Economic Forum. (2022). The Future of CBDCs and Cross-Border Payments. https://www.imf.org https://www.weforum.org

Jack, W. & Suri, T. (2014). *Risk Sharing and Transactions Costs: Kenya's Mobile Money Revolution.* American Economic Review. https://www.aeaweb.org/articles?id=10.1257/aer.104.1.183

Jobin, A., Ienca, M., & Vayena, E. (2019). *The Global Landscape of AI Ethics Guidelines.* Nature Machine Intelligence.

Bibliography & References

Kangas, Olli, et al. "The Basic Income Experiment 2017–2018 in Finland: Preliminary Results." *Reports and Memorandums of the Ministry of Social Affairs and Health 2019:9*, Finnish Government, 2019.https://julkaisut.valtioneuvosto.fi/bitstream/handle/10024/161361/Report_TheBasicIncomeExperiment20172018inFinland.pdf Government Publications+1Government Publications+1

Khan Academy. (2023). *AI-Powered Personalized Learning Experiences*.

Kiva Protocol & Digital Identity – Kiva. (2020). "Sierra Leone Digital Identity Platform". https://www.kiva.org/protocol

Lee, Kai Fu. *AI Superpowers* (The future of AI-driven economies).

Luckin, R., et al. (2018). *AI in Education: The Challenges and Opportunities of AI in Learning.*

Manyika, J., et al. (2017). *Harnessing AI for Business Growth and Efficiency.*

MDPI Media AI Paper. How Generative AI is Transforming Journalism: (2024)

Morozov, E. (2013). *To Save Everything, Click Here: The Folly of Technological Solutionism*. PublicAffairs.

M-Pesa & Financial Inclusion – Jack, W., & Suri, T. (2014). "Risk Sharing and Transactions Costs: Evidence from Kenya's Mobile Money Revolution." *American Economic Review*. https://www.aeaweb.org/articles?id=10.1257/aer.104.1.183

Ng, A., et al. (2020). *AI-Powered Learning: How Artificial Intelligence Improves Skill Development.*

Nguyen, T., et al. (2021). *The Role of AI in Enhancing Accessibility in Education.*

NPR, March 4, 2021 "California Program Giving $500 No-Strings-Attached Stipends Pays Off, Study Finds.". https://www.npr.org/2021/03/04/973653719/california-program-giving-500-no-strings-attached-stipends-pays-off-study-findsNPR

OpenAI. (2023). GPT Limitations & Responsible Use Guidelines.

Pedra. (n.d.). *Real estate virtual staging case studies*. Retrieved March 30, 2026

Patel, R., et al. (2021). *AI-Driven Job Hunting: How Algorithms Improve Resume Visibility.*

Pictory - AI-Powered Video Generator. https://www.pictory.ai

Reuters. (2024). *AI Deepfake Causes Global Panic Before Being Debunked.*

Shapiro, M. (2022). Culture at Risk: The Digital Displacement of Rural Trade. Rural Economics Journal, Vol. 18

Synthesia - AI Video Creation Platform. https://www.synthesia.io

The Future of AI in Content Creation (Harvard Business Review, 2023)

The Silver Generation & the AI Revolution

Bibliography & References

U.S. Federal Reserve Reports on Digital Currencies – Insights into government oversight of crypto (federalreserve.gov).

United Nations. (n.d.). *AI for Good: Leveraging AI to Reach the SDGs*. https://aiforgood.itu.int

Vincent, J. (2022). *Meta's AI Chatbot Makes Dubious Historical Claims*. The Verge.

Wikipedia contributors. (2024). *Everything Everywhere All at Once*

World Bank. (2021). *Global Findex Database: Financial Inclusion and Technology*. https://globalfindex.worldbank.org

World Economic Forum. The Future of Jobs Report, 2020.

Yang, Andrew. The War on Normal People: The Truth About America's Disappearing Jobs and Why Universal Basic Income Is Our Future. Hachette Books, 2018.

YouTube. (2023). *Updated Guidelines on AI-Generated Content and Monetization.*

About the Author

D.M. Lewis has decades of experience, from navigating the complexities of cybersecurity, privacy and risk management to overcoming challenges and personal reinvention.

Her journey is one of resilience, determination, and transformation. With degrees in both Computer Science and Organizational Security & Management, she has built an extensive career at the unique intersection of technology and cybersecurity.

Having worked in environments where financial security, privacy and digital transformation shaped global economies, she has observed the impact of AI-driven opportunities as well as challenges.

Her journey was far from easy. After years of dedication in the corporate world, although many great experiences, she encountered the all-too-familiar barriers of 9-to-5 politics, where at times, talent and hard work often took a backseat to bureaucracy. While navigating an ever-changing landscape of security and risk management with tireless commitment, where breaking through to leadership often required more than just expertise, these challenges fueled her determination to carve her own path, to seek independence, and to create something, **this book,** that could empower others.

Beyond her professional expertise, D.M. Lewis holds globally recognized cybersecurity certifications including:

- Certified Information Security Manager (CISM)
- Certified in Risk and Information Systems Control (CRISC)
- Certified Data Privacy Solutions Engineer (CDPSE)

Her love for technology is balanced with a passion for the simple joys of life, like nature, sharing home-cooked meals, and simply basic acts of charity and kindness. These moments remind her that while AI technology continues to evolve, what truly matters are good human relationships, lessons learnt, and the legacy we build throughout this life on this earth with those we love and care about.

This book reflects her passion to equip others, her readers, various age groups, with knowledge, where they see fit; to break down barriers in this ever-changing financial and technological landscape, and to inspire those seeking control over their future in an AI-driven world. Her story is proof that no matter the challenges, corporate struggles, industry shifts, or personal obstacles there is always a way forward. Look for a way forward.

This book is that way forward.

www.ingramcontent.com/pod-product-compliance
Ingram Content Group UK Ltd.
Pitfield, Milton Keynes, MK11 3LW, UK
UKHW020719041026
15590UKWH00002B/757

9 798995 337904